INTERACTIONS BETWEEN LAND USE AND FLOOD MANAGEMENT IN THE CHI RIVER BASIN

Kittiwet Kuntiyawichai

Thesis committee

Thesis supervisors
Prof. dr. ir. E. Schultz
Professor of Land and Water Development
UNESCO-IHE Institute for Water Education
Delft, the Netherlands

Prof. dr. S. Uhlenbrook
Professor of Hydrology
UNESCO-IHE Institute for Water Education
Delft University of Technology
Delft, the Netherlands

Other members
Prof. dr. ir. R. Uijlenhoet
Wageningen University
Wageningen, the Netherlands

Prof. dr. ir. N.C. van de Giesen
Delft University of Technology
Delft, the Netherlands

Prof. dr. T. Tingsanchali
Nakhon Pathom Rajabhat University
Nakhon Pathom, Thailand

Dr. N.E.M. Asselman
Deltares
Delft, the Netherlands

INTERACTIONS BETWEEN LAND USE AND FLOOD MANAGEMENT IN THE CHI RIVER BASIN

Thesis
submitted in fulfilment of the requirements of
the Academic Board of Wageningen University and
the Academic Board of the UNESCO-IHE Institute for Water Education
for the degree of doctor
to be defended in public
on Friday, 11 May 2012 at 3 p.m.
in Delft, the Netherlands

by

Kittiwet Kuntiyawichai
Born in Buriram, Thailand

CRC Press/Balkema is an imprint of the Taylor & Francis Group, an informa business

Published by
CRC Press/Balkema
PO Box 447, 2300 AK Leiden, the Netherlands
e-mail: Pub.NL@taylorandfrancis.com
www.crcpress.com - www.taylorandfrancis.co.uk - www.ba.balkema.nl
ISBN 978-0-415-63124-2 (Taylor & Francis Group)
ISBN 978-94-6173-249-1 (Wageningen University)

Table of contents

Acknowledgement

It took a long time to complete this PhD research work, even though not nearly as long as it took to build the Royal Grand Palace in Bangkok, Thailand. During my PhD study I always tried out something new, sometimes it worked and sometimes it did not in the end, and of course this was sometimes frustrating. However, I experienced that tracking what I have done helped to overcome the disappointment.

Obviously, the PhD research work is a journey, which takes many years to complete. During this exciting, stimulating, exhausting, long and difficult journey, many people have contributed to this thesis in one way or another as a team effort. I benefited from being supported by a genuinely awesome group of great people who inspired, supported, encouraged and influenced me achieving my tasks in multiple ways. Therefore, I would first like to offer my sincere thanks to a few people explicitly.

First and foremost I want to thank my esteemed promoters Prof. dr. ir. E. Schultz and Prof. dr. S. Uhlenbrook. It has been an honour to be their PhD student. I am especially grateful for them who supported me from the beginning until the end, and been consistently encouraging through all the many ups and downs. Without their constant encouragement, thoughtful guidance, insightful discussions, and valuable advice, I could not have finished this dissertation.

I am greatly indebted to Dr. ir. F.X. Suryadi, for his patience, flexibility, commitment to a high standard, and an ideal combination of offering expert guidance and suggestions while giving me the freedom to follow my own vision, ideas and preferences in a more navigable way.

I would like to gratefully acknowledge the enthusiastic collaboration with Dr. ir. A. van Griensven and Dr. ir. M. Werner. I profited very much from them, and I am thankful for their vast knowledge and experience.

I am indebted to the thesis defence committee Prof. dr. ir. R. Uijlenhoet, Prof. dr. ir. N.C. van de Giesen, Prof. dr. T. Tingsanchali and Dr. N.E.M. Asselman, for the comments and suggestion.

Furthermore, I thank my former Head of the Department of Civil Engineering, Faculty of Engineering, Khon Kaen University, Assoc. Prof. dr. Watcharin Gasaluck, for giving me the chance to continue my study in a PhD programme in the Netherlands. Great appreciation is also given to the Head of the Department of Civil Engineering, Faculty of Engineering, Khon Kaen University, Assoc. Prof. Chinawat Muktabhant, the Head of the Hydraulic Division, Assoc. Prof. Winai Sri-Amporn, Assoc. Prof. dr. Pongsakorn Punrattanasin, for their trust and being very supportive and open minded. Also special thanks to my colleagues in the Department of Civil Engineering, Faculty of Engineering, Khon Kaen University, for their long lasting solidarity.

At this point, I would especially like to recognize my brother Assoc. Prof. dr. Kittisak Kuntiyawichai, I wish to acknowledge his supportive feedback, time and infinite discussions on finite topics. Your suggestions have been essential for solving many problems and for proceeding faster.

Personal thanks particularly go to Dr. Panya Polsan for his hospitality, assistance and good humoured tolerance. Intensive professional guidance during the comprehensive field visit of the Chi River Basin would not have been done so successfully without his assistance.

The following institutes are thanked for their collaboration to provide data to support this research work: Royal Irrigation Department (Thailand), Thai Meteorological Department, Department of Water Resources (Thailand), Land Development Department (Thailand), Electricity Generating Authority of Thailand, Department of Public Works and Town & Country Planning, and Department of Disaster Prevention and Mitigation, Thailand.

My journey to UNESCO-IHE finally reaches its end and UNESCO-IHE must be recognized as a wonderful place for scientific work as well as enjoyable throughout the period of my PhD study. Special thanks to all friends and staff of the institute for providing a stimulating academic environment with their coherent cooperation.

Unforgettable, I propose special thanks to all of my Thai friends in the Netherlands and Thailand, for giving me your support, attention, motivation, kindness, help, solidarity, and company.

Most importantly of all, my heartfelt gratitude goes to my parents, brothers and sister for their unflagging love, support, patience, understanding, and encouragement in every possible way, as well as openness and willingness to share such personal information with me. This dissertation would have been simply impossible without them. Their unconditional love is invaluable to me. I value the time, knowledge and experiences they shared with me. I deeply admire my father who patiently encouraged me for higher education. He works actively to support the family and he has never complained notwithstanding all the hardships in his life. He always teaches me the values of hard work to achieve one's goals, and to always have faith in my own abilities. My mother deserves a special mention for setting a living example, encouraging me and emphasizing the importance of being educated. Of course, she is my idol as she is a successful woman in her life. I have no suitable word that can fully describe her perpetual love, invaluable, indispensable help, and constant support when I encountered difficulties.

To my teachers throughout my education, I would have never been able to succeed without the power, honest feedback, and the inspiration you have given me. You have developed many of my skills and strategies; thank you all for passing on your knowledge, boundless energy and passion.

Lastly, the success of this dissertation would not have been possible without the generous financial support from the Royal Thai Government, which is gratefully acknowledged.

Summary

Floods are one of the major natural disasters that have been causing loss of human life and influence on social and economic development. In many cases nature has shown little respect for man's unwise occupancy of nature's right-of-way in terms of urbanisation and industrialisation by occasionally flooding people's properties and taking their lives. However, floods can also bring important ecological benefits to river systems. Presently, it seems that more and more potential flood disaster areas are developing and the following question arises: How much of those severe flood threats are actually caused by different anthropogenic activities? Often, the causes of floods are:

- deforestation in a river basin;
- changes in land use that link with changes in the water balance dynamics;
- urbanisation and settlement in flood prone areas;
- river regulation, dike construction and channelling;
- increase in water levels as a result of natural or man-made obstructions in the floodway e.g. bridges and weirs;
- dam and dike failure.

Generally, there are two major ways that men have attempted to manage flooding:

- *structural measures*: dams, dikes, channel modifications, bypasses, etc., which are designed to reduce the incidence or extent of flooding, i.e. controlling floodwater by storage, containment, flow modification, or diversion;
- *non-structural measures*: flood forecasting, flood warning, control of floodplain development, flood insurance, evacuation, etc., which are planned to eliminate or mitigate adverse effects of flooding.

In the past, structural flood management measures received priority. However in recent years, non-structural responses are being emphasized to reduce the vulnerability as well as enormous economic, social and human losses, and at the same time minimize the ecological impacts on mitigation measures.

The Chi River Basin is located in the northeast of Thailand. The total area is 4.9 million ha with a population of 6.6 million people. It is located in the tropical monsoon region. The annual rainfall varies from 1,000 - 1,400 mm/year. The main river is the Chi River, which is the longest river in Thailand with a total length of 946 km. The physical characteristics of the Chi River Basin vary significantly, including steep topography in the upstream areas and flat low-lying areas, i.e. the broad, flat floodplains at the downstream part, especially near the confluence with the Mun River that is a major tributary of the Mekong River.

Within the context of demographic dynamics, the population growth rate of the Chi River Basin has increased to approximately 0.6% per year since 1993. Most of the growth has occurred in rural areas, where about 61% of the people live. All of these effects, in turn, have important implications for land use dynamics, which directly affect land demand and utilization. Presently, the majority of the river basin population is engaged in agriculture on the 60% of arable land where 41% is paddy fields. The remaining area is forest (31%), urban (2.9%), waterbodies (2.5%), and others (3.5%).

In view of floods, the Chi River Basin has suffered from a number of major floods over the last few decades, causing fatalities, the displacement of people, extensive physical damage along with enormous economic losses, and large impact on nature. It is obvious that existing flood mitigation measures did not have a significant impact in terms of reducing the flooding and total damage caused by floods. Flood mitigation measures in the Chi River Basin are mainly limited to building of dikes as they are generally used to protect urban areas or agricultural land from relatively frequent floods, particularly in the lower Chi River Basin. These defence structures, built by the Royal Irrigation Department in the 1950's, 1970's, and 1980's, were mostly designed to provide protection against a flood with a 10% annual probability of exceedance.

The main objective of this study was to develop an integrated modelling framework that enables a more holistic approach for flood management, in relation to land use and its changes in the Chi River Basin. There is a growing realisation that an isolated flood management approach may achieve a certain degree of flood mitigation of a certain area. However, it may not be the optimal solution to control future flooding disasters. Therefore, an integrated and optimised approach to the provision of flood mitigation measures, i.e. not several stand-alone approaches, is required. A more integrated approach towards managing the consequences of floods is needed in order to take appropriate actions, i.e. carefully selected points and types of interventions. An integrated approach to flood management will bring in the best mix of both structural and non-structural interventions.

Modelling approaches to flood impact assessment

A flood event is a complex hydrological event. Models do not only help to understand this phenomenon better, but are essential for flood risk assessment of the current situation and to assess the impact of suggested changes in flood prone areas.

Hydrologic modelling

To contribute to a better understanding of hydrologic processes and to synthesize all available data sets (time series and GIS data) in the Chi River Basin, the application of a hydrological model is found to be suitable. However, when confronted with this requirement, the question of course arises which model is best suited to this particular study. To answer the question, the process-based hydrological model, SWAT (Soil and Water Assessment Tool), which is a well-developed, widely used and robust model operating in daily time steps, was chosen.

The SWAT model is best suited for rural areas dominated by agricultural land use, which indeed match the conditions of the Chi River Basin. It also allows detailed inputs for vegetation changes and agricultural practices. It can integrate large amounts of data. Moreover, SWAT was selected to be used in this study because it is an open source model, together with some further advantages such as it is free and editable as it can be adapted to match the particular needs. The model is gaining popularity among users and it has been tested successfully under different geographical and climatic conditions worldwide. Therefore, SWAT was selected to investigate the effects of upstream river basin development activities and modelling the river basin response to changes by flood mitigation measures.

The SWAT model was tested to properly identify and quantify tributary inflows from sub-basins. The daily SWAT simulations covered the period January 1, 2000 - December 31, 2001. The chosen simulation period was based on the historical data, as in this period one of the most devastating floods in Chi River Basin's history happened. A 1.5 years initialization period was used, so the impacts of uncertain initial conditions in the model were minimized. Due to the fact that the SWAT model contains many parameters, which are generally not observable, the model must be calibrated to measured data in order to determine the best or at least a reasonable parameter set. In addition, as no daily time series data are available at the outlet of the Chi River Basin, calibration has been undertaken for Ban Tha Khrai gauged site (E18) within the river basin. During the calibration process, a sensitivity analysis was performed in order to determine the six parameters to which the model results are most sensitive. Afterwards, these six parameters were used to calibrate the model for the period June 1 to October 31, 2001.

With respect to the calibration results at the streamflow station E18, several statistical measures were used to evaluate the simulation accuracy, such as the Nash-Sutcliffe coefficient, Root Mean Square Error, Goodness of fit and Mean Absolute Error. When the 2002 data was used as validation set, it was found that there are remarkable good validation results. The Nash-Sutcliffe coefficient and Goodness of fit values were high (greater than 0.85), which indicates that the SWAT model performance proved to be satisfactory and is able to simulate the discharge reasonably well. Thereafter, the calibrated model parameters were used to simulate streamflow using different rainfall scenarios for the predictions of the tributary inflows with a specified probability of occurrence at all the selected points on the Chi River. These tributary inflows then became a key input into the hydraulic model.

Hydraulic modelling

Capturing the complex interactions between river flow, runoff generation and the area that is prone to flooding is the most important factor for simulating the inundation processes in the floodplains. Therefore, the dominant processes need to be well understood to get insight into the system hydraulics. To address this complex dynamic behaviour, the use of a hydraulic model can make it possible to simulate the areas of complex flow patterns.

A pragmatic approach to deal with flood problems in the Chi River Basin was needed. Therefore, in this study, a significant effort was spent on applying an appropriate hydraulic model, i.e. 1D/2D SOBEK, for the (integral) simulation of flood processes. The utilized 1D/2D SOBEK model has been adopted for the modelling because of its ability to handle flooding as it can dynamically link 1D nodes to 2D cells in the floodplain module. Additionally, it also gives many possibilities for coupling with other models to improve the capability to address key flood management questions.

Model calibration with appropriate data is a crucial step in the overall creation of a valid and reliable process representation. The primary calibration parameter for the 1D/2D SOBEK model is Manning's roughness coefficient (n) of the river channel, which can be determined from the calibration process using 1D modelling (1D SOBEK). The model has been calibrated with observed water levels for the years 2000 to 2001 and adjusted within reasonable limits until the model acceptably reproduced measured water level profiles at gauges downstream of the Chi River. Owing to the fact that the Manning's n values may vary between locations, the model calibration focused on calibrating three sets of Manning's n values. The optimum values were found by comparing the observed and simulated water levels on the Chi River at Ban Tha Khrai (E18) and Maha Chana Chai (E20A) gauged sites. The shape of the stage hydrographs and timing of peaks matched well with the observations, which showed that the model simulates the flood dynamics well as indicated by the high goodness of fit parameters, i.e. goodness of fit is 0.94 at E18 and 0.93 at E20A. The results of the model validation (year 2000) are almost as good, i.e. goodness of fit of 0.79 at E18 and 0.89 at E20A. If one takes into consideration the goodness of fit criterion, the adjusted Manning's roughness coefficients throughout the routing reach give satisfactory results of calibration, which were found to vary from 0.028 s/m$^{1/3}$ in the vicinity of the mouth to 0.045 s/m$^{1/3}$ in the sections further upstream. The values of Manning's n of 0.045 s/m$^{1/3}$ for the routing reach from Ban Non Puai gauged site (E5) until the Chi-Nam Phong confluence, 0.032 s/m$^{1/3}$ from the Chi-Nam Phong confluence until the Chi-Nam Yang confluence, and 0.028 s/m$^{1/3}$ from the Chi-Nam Yang confluence until the Chi-Mun confluence were found to give satisfactory results of calibration and correspond to reasonable Manning's n values according to Chow's table. However, no further calibration could be performed for the hydraulic roughness of the overland flow module (2D SOBEK). The parameter has therefore been derived from the literature (0.1 s/m$^{1/3}$). It was assumed constant throughout the floodplain, as detailed spatial roughness information was not available.

Integrated hydrologic and hydraulic modelling

It is evident that hydrologic and hydraulic processes need to be modelled together through truly integrated modelling by development of a robust and hydrologically sound set of algorithms that are (fully) integrated into a user-friendly interface. As a result, the optimal model is tailored by serially coupling appropriate modules/components for the desired application. Although the model coupling is necessary, it can be quite complicated to implement for many reasons, i.e. data formats, compatibility of scales, etc. However, the openness of the Delft-FEWS platform, i.e. a software developed by Deltares, allows full coupling and data exchange between hydrologic and hydraulic models. This makes it possible to capture the complex interactions, yet relatively simple to apply to various applications. Moreover, as Delft-FEWS is able to integrate the models into one convenient package of flood modelling, interfacing between the two models may be a very efficient way to demonstrate the versatility of this tool for flood inundation studies.

In this study, the rainfall-runoff hydrological model was used to generate the runoff deriving from two 'critical events', the first based on the 2001 flood, which corresponded to a 4% annual probability of exceedance flood and the second based on the extreme rainfall with a 1% annual probability of exceedance. The hydrological simulations provide inputs to the following hydraulic model to simulate the floodplain inundation in regard to the detrimental impact of potential flooding. The coupling of the two models was carried out in a multi-scenario flood modelling experiment, i.e. analysing future changes in hydrological and associated systems, land use, etc.

The coupling between SWAT and 1D/2D SOBEK was made through a series of nodes where tributaries connect to the Chi River, and it had to be assumed that there is no direct feedback of the overland flow to the rainfall-runoff response. The coupled SWAT-1D/2D SOBEK model performs as follows:

- rainfall-runoff module (SWAT) is the starting point of the coupled model to give a realistic representation of the terrestrial hydrological systems as it forms a link between three main hydrologic compartments: atmosphere, surface water, and groundwater. The hydrologic inputs define the magnitude of total stormflow from the various sub-basins;
- thereafter, flows at the outlet of the sub-basins serve as inflow boundaries to the overland flow module (1D/2D SOBEK) at specified locations by coupling nodes in the river network. At this step, the SWAT simulations for flow routing in reaches on the main stem of the Chi River were turned off. Subsequently, the ensuing flood propagation is simulated in a half-hourly time step even though the input data used

from the SWAT simulations were at a daily time step as the 1D/2D SOBEK model is capable of interpolating input data for smaller time intervals. The way of interpolation may influence significantly the performance of the model. A high mass error occurs if the time step is too large as the interpolating input data changes suddenly from one value to a different value, i.e. having a greater peak to peak variation at each time step and causing numerical problems. However, due to the relatively small time step used in 1D/2D SOBEK, no significant effect on the hydraulic simulation could be found.

The combination of the two widely used models SWAT and 1D/2D SOBEK is able to simulate flood generation processes realistically and to assess the impact of various flood mitigation scenarios, i.e. both structural and non-structural measures, on various factors related to floods, e.g. increased runoff volume and flashiness, increased flow retardation, etc. Regardless of the end result, model coupling has proven to be feasible and efficient as it seems to be a promising approach with significant benefit to flood management in the Chi River Basin. Further, while designed for the Chi River Basin, this model coupling may be used as a prototype for model applications in other areas of the country.

Understanding the influence and impact of uncertainty in the context of flood modelling is crucial for flood risk management. To improve the flood modelling process, it is crucial to further understand the uncertainties inherent in the modelling processes, and how the uncertainty is translated to the model results explicitly, which may be included in further studies.

General analysis of alternative flood mitigation measures

At a general level, the benefits and costs of a certain (set of) measure(s) have been indicated. For an alternative flood mitigation measure, the flood damage reduction (benefit) would have to exceed the implementation cost. In comparing alternatives, the net benefits, i.e. total benefits minus total costs, would have to be greater than in other alternatives that achieve the same flood reduction.

Flood damage quantification

The damage caused by floods can be approached as a function of the flood characteristics, i.e. depth and duration of flood inundation, due to physical contact with floodwater per category of element at risk. In Thailand, structural measures are designed to withstand the effects of floodwaters, which are caused by an event with a given probability of exceedance. According to the Royal Irrigation Department, the exceedance frequency (safety standard) has been fixed at 1% annual probability of exceedance at a given location where many people might be potentially affected. Therefore, in this study, the damage potential is assessed on the basis of the calculated flood depth with a 1% annual probability of exceedance for riverine flood events in order to indicate the vulnerability to inundation, and to show the spatial distribution of potential damage across the Chi River Basin. As a result, the concerned values were calculated in order to estimate the direct benefits of flood mitigation measures in terms of flood damage reduction. Note that impacts such as human health, environmental damage and benefits or other indirect costs were not considered in this study.

Spatial analysis techniques enable integration of flood depth and land use to evaluate which elements or assets are affected by a flood with a 1% annual probability of exceedance, and how much they are affected in terms of inundation depth. The following land use categories were considered in the damage assessment: residential, commercial, industrial, agriculture and infrastructure.

Damage functions were adopted for the quantification of different damage categories in monetary terms. Based on land use, asset values and damage functions, direct damage caused by the flood with a 1% annual probability of exceedance was determined. However, direct damage to infrastructure was not taken into account. Therefore, in this study, the damage to infrastructure was estimated as a fixed 65% fraction of the total damage of the flood losses. Using such damage functions, damage to different land use categories was estimated and the summation indicated the total direct flood damage.

Estimated implementation costs for flood management alternatives

The estimated implementation cost for each alternative was estimated from the sum of the costs associated with construction, operation and maintenance. The construction cost was calculated based on the Thai Bureau of the Budget Handbook, based on April 2009 unit rates (note: these costs are rough estimates). Besides the construction cost itself, this estimate also included the costs related to the operation and maintenance. It was supposed that these costs represent annually about 5% of the construction cost in order to serve the primary purposes of interventions during the designed lifespan, which generally concerns 50 years. The expected damage has also been determined based on this period. Furthermore, it needs to be noted that the estimated implementation cost does not cover the cost of land acquisition where envisaged flood mitigation measures are to be located. Therefore, the cost presented would have to be considered as indicative only and detailed investigations are necessary to obtain more accurate cost estimates. These will be probably higher. Due to this, at the present stage, the focus has been on the land use, hydrologic and hydraulic aspects and the cost benefit analysis has a preliminary character.

Optimum level of flood mitigation

The effects of flooding can be mitigated, and thereby the loss of life and damage to property can be reduced. Adoption of a certain flood mitigation alternative depends on the hydrological and hydraulic characteristics of the river system. However, flood mitigation measures cannot be evaluated from a single point of view. The technical performance of these measures, in terms of preventing inundation and the resulting damage, needs to be taken into consideration as it is important for an overall appraisal of the acceptability of each alternative.

The optimum level of flood mitigation cannot eliminate all flood risk. Realistically, it can be expected to only minimize the total flood mitigation costs and damage due to residual flooding. It refers to the point where the sum of construction, operation and maintenance costs and damage are minimized for each flood mitigation alternative.

Identification and appraisal of the feasibility of flood management interventions

The Chi River Basin has been subject to flooding and increased flood risk to people and assets because of physical and operational constraints of the existing flood management systems, reliance on flood management facilities that do not provide the level of protection up to the required level, changing land uses in flood prone areas and limited understanding of flood risk. However, reliable flood alleviation schemes would need to ensure a realistic minimum disruption from flooding, and are subject to local acceptance. The efficiency of flood management can be achieved by ensuring that appropriate actions in response to flood mitigation are timely undertaken with more focus on the type and location of measures. This study provides an inventory and description of various options for flood management in the Chi River Basin. However, only the most common measures recommended today that strengthen an integrated approach to flood management are discussed. These alternatives were derived through the preliminary understanding of the physical situation, analysis of flood behaviour, identification of needs, and review of previous studies. The alternative measures have to ultimately meet the following evaluation criteria:

- feasibility, with a view to incorporating anticipated flood damage;
- technical effectiveness, in view of effectiveness in reducing flood extent.

Prior to designating the possible flood mitigation measures, the 1D/2D SOBEK model has been applied to determine which areas might be prone to flooding from a flood with a 1% annual probability of exceedance, and where the proposed options might be put in place that will not cause adverse impacts on existing flooding conditions downstream.

Towards inundation simulation based on a flood disaster scenario with a 1% annual probability of exceedance, four possible flood mitigation alternatives were analysed to estimate their hydraulic effectiveness in reducing flooding of critical sites in the Chi River Basin. Potential hydraulic impacts of flood mitigation were assessed by comparing pre and post mitigation flood depths and inundation extents. In order to obtain the preferred option, indicative cost-benefit analyses have been applied. As a result, the optimal performance of the preferred option is found by minimizing the resulting total costs.

River normalisation

Flood disasters are caused by discharge that is higher than the capacity of the existing river causing water to overflow the adjacent lands. The river capacity is often reduced because of the narrowing due to extensive vegetative growth, accumulation of debris, etc. This is the reason why river normalisation can be a viable option; in some cases it can be done by reducing hydraulic roughness, increasing cross-sectional area, or reducing potential for blockages and hang-ups of drift.

An estimate has been made in order to identify potential effects of river normalisation on stream function, which will change the flood runoff of the river. Therefore, river normalisation has then been modelled by changing the hydraulic variable, i.e. Manning's roughness coefficients of the river bed, which has been determined from the calibration process, in the 1D/2D SOBEK model.

Seven scenarios for river normalisation were defined and analysed, which correspond to the number of reaches throughout the routing reach (i.e. Ban Non Puai gauged site (E5) until the Chi-Mun confluence). If river normalisation of the Chi River was to be considered in more detail then the following actions would be required:

- make a preliminary assessment of the costs of normalisation;
- quantify the benefits that are derived by normalising the Chi River to mitigate flooding during the flood with a 1% annual probability of exceedance.

The 1D/2D SOBEK model of the Chi River has shown that removing hydraulic obstacles, vegetation, and bank clearance in the lower reaches of the river, i.e. normalising the reaches from the Chi-Lam Pao confluence to the Chi-Mun confluence, would have a significant effect on flood characteristics in the floodplain. The Manning's n roughness coefficients of the lower reaches are currently set at 0.032 (from the Chi-Lam Pao confluence to the Chi-Nam Yang confluence) and 0.028 (from the Chi-Nam Yang confluence to the Chi-Mun confluence) in the hydraulic model. Assuming that river normalisation reduces these roughness coefficients to 0.025 then the inundated area with a 1% annual probability of exceedance will decrease by approximately 16,000 ha. The cost of damage associated with flooding (both to physical properties and agricultural land) is expected to decrease from US$ 86 million to US$ 64 million across the river basin. The results show that this prevented damage is regarded as a benefit of river normalisation that exceeds the implementation cost, which has an estimated cost of US$ 21 million.

It is necessary to pay careful attention to the selection of methods of river improvement as it has some important shortcomings that need to be kept in mind. It should not disrupt the existing morphological balance of the river system. To compensate for the alteration in the hydraulic variable and to establish a new/stable equilibrium, other parameters will change. Therefore, river improvement will probably have a great impact on a river because it can disrupt the existing physical equilibrium of the river system and the potentially large environmental effects (less habitats, impact on groundwater recharge, etc.), unless the operations are carefully planned, conducted, and monitored. Also, regular maintenance must be carried out to ensure continued satisfactory operation.

Reservoir operation

Flood control reservoirs can contribute significantly to flood management as they can function effectively to capture and temporarily hold floodwater upstream of the flood liable areas, and release it gradually later. Large areas of land required to locate a new reservoir are no longer available, especially where it would affect large populations and good agricultural lands.

The contribution of the Nam Phong and Lam Pao Rivers, where the Ubol Ratana and Lam Pao reservoirs are already located, towards reduction of the flooding of the downstream area of the Chi River Basin was studied. The modelled scenarios were explored by changing initial reservoir volume and daily discharges from the existing reservoirs at the beginning of the flood season, to reduce the impacts of a flood with a 1% annual probability of exceedance. The optimum design and operation conditions, i.e. the percentage of initial reservoir volume and reservoir outflows, were achieved by applying the water balance equation calculated from inflow, outflow, losses, and reservoir storage. Moreover, in the reservoir releases the downstream needs for water supply, agriculture, and environment, requirements for flood regulation, storage considerations, and legal requirements for minimum flows have also been taken into account.

Simulation techniques are not able to generate directly an optimal solution to a reservoir operation alternative. However, by going through several model runs with alternative operating rules, near optimal operating solutions can be identified. In this study, five scenarios of reservoir operation have been compared to identify the most efficient type of operation on reduction of potential flood damage. The results revealed that the scenario in which during the flood season, the outflows released from Ubol

Ratana and Lam Pao reservoirs are about 55% and 75% of their original daily outflows, respectively, while the reservoir volumes are about 50% and 30% of their initial volume at the beginning of the wet monsoon, respectively, gives the best results. By following the proposed reservoir operation rules, there is a reduction of about 14,100 ha in flooded area and the potential flood damage is expected to decrease to US$ 70 million from US$ 86 million. It needs to be noted that the reservoir operation alternative has no implementation costs while only the reservoir operation rules would have to be adapted. However, operation of reservoirs needs very careful consideration in further studies, with special attention paid to direct and indirect damages, including reduction in benefits from the reservoir in terms of hydropower generation, irrigation, water supply, tourism, etc., which probably make the reservoir operation alternative expensive. However, both reservoirs can significantly increase the amount of storage space to store impending floods up to their storage capacity until the end of the flood season. Moreover, one needs to bear in mind that the overestimation of the initial reservoir volume and reservoir outflows can lead to substantial losses. In addition, reservoir operation practice alone typically cannot be considered a reliable alternative to safely pass tremendous floods. Therefore, this alternative would have to be integrated with other flood mitigation measures, e.g. through legislation and controlling the developments in the floodplains.

Green river (bypass channel)

In the Chi River Basin, the main floodplain is regularly flooded, which is caused by overtopping of low-lying points on the river banks of the Chi River. To mitigate overbank flooding, a potential solution consists of a set of green river channels, in order to keep the flow within the capacity of Chi River and lessen the impact of flooding on the main river system (note: it is called 'Green river' since it can provide a natural green space such as park or agricultural area after the flood peak in dry periods. In different circumstances, water can also be stored in this zone for agricultural purposes).

To designate a green river channel, hydraulic modelling has been carried out to determine the area of flooding that would result from a flood corresponding to the 1% annual probability of exceedance. The designated green river channel is defined as the channel and those parts of the floodplains adjoining the channel, which are reasonably required to provide for the passage of flood. Seven possible scenarios have been analysed and optimised with the 1D/2D SOBEK model for selection of the most promising scenario. It has been found that the alternative consisting of a designated green river channel system that includes two green river channels covering approximately 53 km of channel length on the right bank floodplain of the Chi River is the most effective. The part of the flow greater than the bankfull capacity of the Chi River is diverted into the green river channels. The first spill from the Chi River occurs in the vicinity of Chi-Lam Pao confluence through a 36 km-long green river channel. Moreover, floodwater once again leaves the river through another green river, which would be in place at the most downstream end of the Chi River over a 17 km length. The preliminary design of the green river channel has been made based on a channel that would be trapezoidal in shape with a bed width of 25 m, 1:1 side slopes, and 3 m depth. The control would be in the form of spillways or gates at the entrance to the green river. With these green river channels, the flooded area of the 1% annual probability of exceedance flood can be reduced by 8,000 ha. Moreover, the potential benefit created by this measure, which is defined as the decrease of damage cost, would be about US$ 10.8 million. As the implementation cost for this alternative is estimated to be about US$ 9.3 million, therefore, the counter measure may still be beneficial after expenses are considered without considering the costs of the land for the green channels. However, land uses within an adopted designated green river channels must be restricted to not impede the free flow of floodwater or endanger public safety. In general, activities such as agriculture, grazing, and wetland are allowed if a good flood warning system is in place, as they can be quickly and easily removed or pose little impedance to river flow.

Retention basins

In response to the inevitable flooding, the provision of off-channel retention storage is found to be a possible action for consequential reductions in overbank flooding and associated damage. The retention basin will simply allow the storage area to be filled under controlled conditions when water levels or flow rates exceed a certain limit. A control structure, which operates based on pre-defined conditions, will be used to hold part of floodwater in the storage area in order to reduce the flood depths and flood extents further downstream. The floodwater in the storage is then released when the Chi River can carry the flow without the water level exceeding the levels of the banks. However, since the desired effect of a retention basin occurs downstream of the Chi River Basin, a location as far upstream as possible, while still being located such at substantial amounts of water can be stored, is preferred.

In this study, the identification of four potential sites situated alongside the Chi River has been taken into consideration at the preliminary phase; subsequently, the most promising one has been selected for a more detailed analysis. The results revealed that a retention basin in the upper Chi River Basin can provide a significant downstream damping effect. For this reason, the proposed construction of 5,500 ha upstream floodwater retention on the left bank floodplain of the Chi River is located between Ban Khai and Ban Kaeng Ko gauged sites, which are typically considered as repeatedly flooded areas. To achieve the required storage, off-channel retention storage would require the construction of a dike with a maximum height of 9 m, trapezoidal-shaped flood diversion channels (i.e. 25 m width at the base, 1:1 side slopes, and 3 m depth), and control structures to enable particular areas of the floodplain to be inundated to specified depths at particular times. By routing the design flood through the potential retention basin, there was a reduction of about 5,000 ha in the flooded area. A cost-benefit consideration showed that for an investment of about US$ 17.0 million, selected application of the construction of a retention basin would yield benefits up to US$ 17.3 million by reducing flood damage.

Selection of alternative measures

To guide which set of flood mitigation measures would have to be considered in this study, the alternative measures were put through a screening process based on a decision matrix approach. The analysis of the above results reveals that the alternatives can meet the screening criteria, which is indeed promising enough to be considered further. However, since single alternatives sometimes induce effects opposite to others, a suitable combination might be desirable in order to reduce the overall risk.

Optimal combination of flood mitigation measures

It is not possible to prevent or even reduce flooding everywhere. Therefore, it is necessary to determine the most effective use of resources and expenditure. In the context of growing interest in finding appropriate solutions to flood management in the Chi River Basin, effective responses may involve a set or judicious combination of flood mitigation approaches rather than reliance on a single measure. In this study, from the best case scenario for each alternative, a series of scenarios have been formulated, i.e. a total of 11 possible combinations/scenarios were analysed as follows (note: the total implementation costs would vary in each scenario):

- Scenario I-VI: combination of two alternatives;
- Scenario VII-X: combination of three alternatives;
- Scenario XI: combination of all four alternatives.

Through detailed hydraulic modelling, the best-case scenario has been identified by means of a set of priority scoring criteria in consideration of technical and financial efficiency aspects. The investigations revealed that Scenario XI would be the optimal one, which includes the following major measures:

- approximately 331 km of river normalisation;
- changes in operation rules of the Ubol Ratana and Lam Pao reservoirs;
- two green river channels, in total 53 km long;
- 5,500 ha off-channel retention storage.

From technical as well as financial point of view, the optimal combination would reduce more damage compared to one of the singular measures; it may reduce US$ 48 million of the flood damage. In the present approach the potential damage, which can be obviated by implementing the optimum combination of flood mitigation measures, is greater than the total implementation costs, i.e. US$ 47 million. In this respect, it is remarkable that an optimal solution would enable the inundated area to be decreased from approximately 143,000 ha to 101,000 ha for a flood with a 1% annual probability of exceedance.

Interactions of land use planning and floods

A substantial area of the Chi River Basin has been changed mainly due to increased pressure resulting from human interventions for settlement and agricultural expansion, infrastructure development, deforestation, etc. The land use change can affect the hydrological behaviour of a river basin through the influences of land uses on the runoff generation processes. These changes may alter the quantity of surface/subsurface runoff generation, river flooding regimes and extent. To mitigate flooding susceptibility in the Chi River Basin, the Royal Thai Government will enable a significant change to adopt better management practices in agriculture, forestry, land use planning, water resources management, and urbanization. To quantitatively reveal this effect for the entire river basin, an evaluation of land use change was analysed first.

The development of potential river basin change scenarios will reflect possible futures looking 50 years ahead. Whilst it is not possible to perceive what will occur over the next 50 years, future trends can be projected to determine the scale of change that would affect flood risk. According to the government land use planning for the year 2057, the land use distributions are predicted to undergo changes dominated by afforestation. The forest proportion of the total river basin area is predicted to increase from 17.9% to 37.2% from 2000 - 2002 to 2057. The gains in forested areas result in losses in all non-forest land uses. Agricultural areas (75.5% of the total river basin area) will have the largest area losses (up to 18.9% in 2057). The urban areas will decrease 0.9% and this decrease will be replaced by the industry for approximately 0.5%.

The individual model runs have demonstrated the potential of the hydrological model for assessing the impacts of land use changes on the hydrological regime of the Chi River Basin. The simulation results indicate that the future changes in land use will result in a significant decrease in flood flows. It can markedly reduce the threats of flood hazards in the upper areas of the Chi River Basin compared to current land use if a systematic and realistic approach to forest land use planning in the upstream part of Chi River Basin is applied. For instance, the reduction of the (1% annual probability of exceedance) peak discharge can be as much as to approximately 10% at the outlet of one of the upstream sub-basin. However, the differences in peak flow and discharge volume varied strongly from one sub-basin to another.

The analysis of future land use change in the Chi River Basin found that many of the sites, which are currently affected by flooding, seem not likely to experience significantly changed flooding in the future. In other words, adjusting the land use may not have a significant mitigating impact on reducing flood flows as the reductions for the downstream sub-basin outlets are small. An underlying assumption is that the climate remains stable in this region and all other possible environmental changes are considered in the modelling approach.

To better understand how changes in land use may affect future flood risk, the 1D/2D SOBEK model was used to test future changes against current baseline results. The results showed that the overall area of flooding decreases only insignificantly across the river basin as a whole, i.e. from 143,000 ha to 142,000 ha, during a flood with a 1% annual probability of exceedance. The decrease in flood extent is therefore marginal under the future land use scenario. Although the flood extent is likely to decrease slightly in future, costs of flood damage will continue to increase significantly due to the most remarkable change in the number of commercial properties at risk of flooding. Under the future land use scenario, the cost of damages associated with flooding is expected to rise from US$ 86 million to US$ 140 million across the river basin. Minimizing the increase in the cost of flood damage, a scenario was used to test the effectiveness of optimum combination of flood mitigation measures against a range of likely futures. The results suggest that the use of the chosen optimal solution can help to reduce future flood damage from US$ 140 million to US$ 85 million, and will reduce the flooded area by another 39,000 ha.

For the reliability of the model simulated impacts of land use change, it is essential to filter out as much land use planning data uncertainty as possible. In real world applications, uncertainty always exists in the model outputs as a result of uncertainty concerning the spatial allocation of predicted land use changes in addition to the general model uncertainties coming from the other input data, the model structure and parameters. However, due to the fact that uncertainty analysis on land use change is difficult to assess and can be attributed to a number of sources, reducing this uncertainty needs to be addressed in further research.

Final notes

The Chi River Basin constitutes approximately one-third of landmass and population of the Northeast region of Thailand. In a view of geography, the landscape is basically a rolling upland with unfavourable agricultural conditions characterized by infertile sandy soils except in the floodplains. The rainfall regime is quite irregular and unreliable, which makes the area vulnerable to droughts and floods. Its past development was mainly dominated by expansion of direct exploitation of natural resources. Farming systems, which includes full utilization of land and water resources, lead inevitably to land and water resource degradation and this becomes more widespread.

Due to the population growth in the Chi River Basin, there has been an immense degree of deforestation. Between 1952 and 2003 the forest cover has dropped by 20%, whereas agricultural and urban areas have expanded rapidly. Typically, the area is characterized mainly by agricultural land use, about 78% of the households engage in agriculture, and are heavily dependent on a few crops, i.e. rice, along with cassava, sugarcane, corn, soybean, peanut, and kenaf. However, current practices entail soil erosion, salinity, and loss of fertility, resulting from expanding the planted areas of cassava, sugarcane and corn, all of which appear to be site-specific. Obviously, it can be seen that crops nowadays are more diversified than what used to be grown, and they are increasingly being grown on suitable land in the area, partly because of achievements in agriculture development. Presently, there is approximately 0.54 million ha or about 11% of the total cultivated area receiving water from 1,836 large, medium and small water resource development schemes, including electrical pumping schemes. Moreover, industry is also increasingly building importance as it mainly provides support services that are an essential part of agricultural production, which includes rice, tapioca, and sugar mills. As a consequence of the development activities in the Chi River Basin, it is recognized that its economy has expanded greatly in recent decades.

To a certain extent, past developments do not necessarily reflect the future. Therefore, the anticipated future impact of activities would have to be accommodated as it would probably cause a significant adverse impact on river basin functions and behaviours. In the future, the continuous growth and development in the Chi River Basin will still be based upon the exploitation of natural resources. However, it needs to deal with care to maintain harmony among sustainable resources and socio-economic demands, as well as to ensure that the productivity of primary natural resources will not deteriorate as a result of development activities in the river basin. In dealing with land resources, land use optimization is found to be essential for the achievement of economic and social benefits. In particular in the agricultural sector, paradigm shifts from extensive to intensive agricultural systems, as well as from dependence on a few cash crops to more diverse crop options in conjunction with more efficient utilization of limited resources, are required. In line with the agricultural strategy, the establishment and upgrade of irrigation facilities to be properly functioning and well-suited to local conditions are also important with a view to enhancing water productivity and contributing to long-term sustainability of agriculture. In sum, the essentials of comprehensive planning need to include the necessary anticipation as much as possible, together with both detrimental side effects and unaccounted benefits from development activities, and to find ways to maximize benefits and relieve the adverse effects through effective management.

Whenever river basin development occurs, it is crucial to be aware of in which way development is associated with the possibility of flood generation, in the sense that development might affect the flooding process and vice-versa. The changes in land use associated with river basin development would have to be considered as they may concentrate and accelerate flows and lead to more excessive damage on existing infrastructure. The specific information relevant in this context are the inundated area and the costs of flood damages during a 1% annual probability of exceedance. Based on the present land use condition, the derived result shows that an area of 143,000 ha will be inundated with an estimated damage of US$ 86 million. The damage potential might increase to US$ 140 million or 1.6-fold as a result of changes in land development patterns in the period up to 2057.

In an attempt to improve the functioning of the river basin, it is therefore necessary to pay particular attention to balance development requirements and losses related to floods. Therefore, a wide range of potential flood mitigation measures has been examined. The screening results indicated that the most cost-effective plan would likely to be the combination of river normalisation, reservoir operation, green river channels, and retention basin. Once they are applied together based on the expected flood extent for a 1% annual probability of exceedance, flood damages (current land use patterns) are likely to decrease in the order of US$ 48 million and thus saving 42,000 ha from inundation. Future flood damage reduction is projected to reach US$ 54 million (39,000 ha reduction in inundated areas), if land use change occurs as currently predicted (by 2057).

On top of the 1% annual probability of exceedance flood, it is worth mentioning that this magnitude of flood can be exceeded. This would mean that anticipation becomes crucial for a further assessment of future flood risk. As far as this issue is concerned, an insight into an exceptionally large flood (e.g. 0.1% annual probability of exceedance flood) is required to place the higher level of safety. If such a flood occurs, under current land use conditions, 165,000 ha of land could be inundated and could cause up to US$ 10 million in direct damage. When the predicted land use changes take place over extended timescales towards 2057, the extent of inundated area is almost equivalent to the case of existing land use conditions, but its damage may increase by as much as US$ 18 million. In reacting to the threats posed by this intensified flood alongside the integrated solutions for flood mitigation, if the present land use trends continue, flood inundation and damage are estimated to reduce by 34,000 ha and US$ 4 million, respectively. Meanwhile, an estimated 32,000 ha in inundation reduction and US$ 6 million in lessening potential damage are expected with the 2057 land use conditions. To this end, it is clear that the optimal combination of measures would significantly reduce flood damages and flood risk in the Chi River Basin, however, the effects of flooding still cannot be completely eliminated.

In summary, the integrated modelling framework that couples the hydrological (SWAT) and hydraulic (1D/2D SOBEK) models has shown that it can be used for much more than detecting, determining, and estimating the flood extent, damage and impacts. Particularly, it can be used to investigate the impacts of land use changes in the entire river basin. Based on the above statements, a good understanding of the relative scale and direction of future changes will be obtained, which can help to minimize future flood risk and damage costs, and this can be used as the best available information to improve flood management.

1 Introduction

1.1 General

Floods are natural events that have always been an integral part of the (geologic) history of earth, and they occur as follows:

- along rivers, streams and lakes;
- in coastal areas;
- on alluvial fans;
- in ground-failure areas such as subsidence;
- in areas influenced by structural measures or their failure;
- in areas that flood due to surface runoff and locally inadequate drainage.

A considerable part of human history has been spent drying land after a flood disaster occurs. In other words, a flood disaster is the process in which flood acts on human livelihood. In fact, floods not only have negative effects, but also positive consequences. Unfortunately, emphasis is often evident in their destructive aspect which are among the most frequent and one of the most costly categories of natural disasters, i.e. one third of the annual natural disasters and economic losses, and more than half of all victims are flood related (Douben, 2006). A series of disastrous floods have led to suggestions that the risk of flooding and the scale of damage will increase in the future. In addition, a number of possible causes are suggested, i.e. building in floodplains, alterations to river systems, changes in rainfall patterns and changes in land management practice, etc. This briefing examines how the response might be improved.

Initially, it is important to distinguish between floods and flooding. These terms are often used inconsistently, thus in this study the terms are defined as follows:

- a *flood* is a temporary condition of surface water (e.g. river, lake, or sea), in which the water level and/or discharge exceed a certain value, thereby escaping from their normal confines. However, this does not necessarily result in flooding (Munich Reinsurance Company, 1997);
- *flooding* is defined as the overflowing or failing of the normal confines of a river, stream, lake, canal, or sea. Also, accumulation of water as a result of heavy rainfall or other events such as dam breaks, by lacking or exceedance of the discharge capacity of drains in the affected areas which are normally not submerged (Douben and Ratnayake, 2006).

After a catastrophic flood event, it is always claimed that influence of human-induced changes in land use have increased its severity. Human settlements and activities have always tended to use floodplains. Their use has frequently interfered with the natural floodplain processes, causing inconvenience and catastrophe to humans (Mays, 2005). Undoubtedly, the greatest dilemma that humanity faces today is a consequence of deterioration of the existing balance of rainfall-runoff in the river basins in favour of flow owing to various reasons, i.e. deforestation, unstable land forms, etc., which make the flood prone areas more vulnerable for floods. As a result, the risk of flooding may change over time due to changing development conditions within the river basin. However, people will naturally attempt to conquer the associated risks following an adverse event, even though sometimes well-intentioned efforts can make matters worse. They choose to alleviate the threats posed by nature through judicious management of water resources and threats in order to manage floods where they arise, not just where they have their effect, with an emphasis on engineering solutions. Over time, structural and non-structural mitigation options are being recognized in order to obtain full benefits and achieve desired results.

Generally, there are two major ways that men have attempted to manage flooding:

- *structural measures*: dams, dikes, channel modifications, bypasses, etc., which are designed to reduce the incidence or extent of flooding, i.e. controlling floodwater by storage, containment or flow modification or diversion;
- *non-structural measures*: flood forecasting, flood warning, control of floodplain development, flood insurance, evacuation, etc., which are planned to eliminate or mitigate adverse effects of flooding.

Single flood mitigation options can only benefit specific areas, but on the other hand, they are not appropriate in many areas, i.e. flood damage may exceed what would have occurred if the option had not been implemented. Therefore, an integrated approach is increasingly endorsed as a crucial support for proactive mitigation efforts, which is far more cost-effective than paying to clean up and rebuild after a flood occurs. In practice, detailed assessments are conducted to quantify the effectiveness of a promising alternative and estimate the associated costs, through a comprehensive hydrological and hydraulic analysis across the entire river system (not just in the areas that might benefit). Consequently, an optimal set of mitigation measures will provide a more robust solution to deal with the increasing development pressure inside flood prone areas and the uncertainties created by changes in land use.

Given its importance, it is likely that the study entitled 'Interactions between land use and flood management in the Chi River Basin' can give a response to challenge the upcoming changes in flood regime and land use.

1.2 Problem description

Floods are one of the major natural disasters that have been causing the greatest loss of human life and influence on social and economic development. The statement became even clearer with the experience of Thailand's worst floods (2011) in 70 years, which had overwhelmed nearly one-third of the country such as the northern, north-eastern and central parts of Thailand. The floods caused substantial damage to the industrial estates, drowning farmland, and tragic loss of hundreds of lives (Bank of Thailand, 2012). Over many centuries, people have influenced the use and extent of natural floodplains. Obviously, human induced changes in land use are one of the significant reasons that exacerbate flood occurrence and its severity in various river basins (Myers, 1997; Schultz, 2001, 2006b). The developments within flood hazard areas can increase the severity and frequency of flooding by reducing flood storage, increasing storm water runoff and obstructing the movement of floodwater. In addition, structures that are improperly built in flood hazard areas may be subject to flood damage and threaten the health, safety and welfare of those who use them. Therefore, in areas where land is heavily used, water would have to be matched to the existing use of space and be combined with housing, industry, commerce, infrastructure, transport, agriculture, etc.

The study area, Chi River Basin in the northeast of Thailand, has always been subjected to problems of flooding, which are getting more severe. The meteorological aspect of high intensity rainfall events, i.e. from tropical storms and depressions, rarely by typhoons, over the area often occurs, causing flash flood from the upstream and stagnant flood at the downstream part. Together with the rising waters of the Mun River that occurs almost at the same time, it has resulted in decrease in drainage efficiency of the Chi River. Consequently, floodwater is accumulated in the vicinity of Chi-Mun confluence, and correspondingly connected to the flood area overflow from Mun River. In addition, because of the flat topography of the downstream part, especially the area of Khueang Nai District, Ubon Ratchathani, the Chi River became extremely meandering, which causes the river to flow in a narrow channel. In case of severe flooding, the floodway will become as wide as the floodplain itself. Moreover, floods can also occur as a result of simultaneous discharge peaks from the Chi River and its main tributaries, i.e. Nam Phong, Lam Pao, and Nam Yang. Another cause of flash and stagnant floods is the reduction of flood discharge capacity in the Chi River as a result of vegetation growth in the channel, accumulation of sediment, snags or debris, additional downstream restrictions, etc. In association with improper land use in the floodplain area, including deforestation, which might result in a higher flood level, longer flood period, and eventually higher flood damage, as this has worsened considerably over the last two decades.

Flooding in the Chi River Basin has long been a recurrent problem (2 - 3 times a year), based on the historical data, significant flooding appears to occur every 2 to 3 years.

Due to the risk of large-scale damage to public and private property, the flood protection infrastructure, i.e. dikes, was built, particularly in the lower Chi River Basin. Apparently, it has only transferred flood damage from one location to another and not solved the problem entirely. Therefore, a comprehensive flood management plan would have to be prepared, aiming at providing effective and adequate flood mitigation to the Chi River Basin for the consequences of flood hazards, response and recovery from flood events.

1.3 Objective and methodology

1.3.1 Research questions

To address the issues detailed in the problem description, an answer to the following main research question with respect to flood management in the Chi River Basin is needed:

'How will the ongoing changes in land use and water management influence floods and flooding, and what would be the implications for flood management?'

It is obvious that despite all the flood mitigation measures taken and the money spent flooding vulnerability and susceptibility in the Chi River Basin are still increasing. Therefore, it is a challenge for this research to focus on the applicability of flood management in combination with land use interactions to mitigate up to a certain chance of occurrence flooding in the Chi River Basin (Kuntiyawichai et al., 2011a; Kuntiyawichai et al., 2011b). Nine specific research questions have been formulated to guide the research. Providing answers to these research questions will permit the main research question to be answered with confidence:

- what are the nature and extent of the local problems with respect to flooding?
- which areas are most susceptible to face present and future flood conditions?
- what flood management measures are currently in place?
- how can flooding be managed when it is beyond the design standards of protection?
- which actions could best be taken and where and when this could be done best?
- what are the possible mitigation measures that can be taken to reduce flood hazards?

- how can the ability of possible flood management measures contribute to increased resilience in response to future flood conditions be assessed/evaluated?
- how to identify the most appropriate flood management strategy under the local conditions in the Chi River Basin?
- how may land use change scenarios affect the chance of flooding, flood damage and the potential for flood management in the Chi River Basin?

1.3.2 Scope of the thesis

In order to limit the research area to an in-depth manner, more relevant, easier to comprehend, and the findings easier to apply, the scope has to be defined and limited as follows.

This study covers the entire Chi River Basin, including all major sub-basins. The view expressed herein does not address water quality, mainly to issues of water quantity, i.e. it is constrained to the effects of flooding from the river system. Within the scope of flood management, consideration will be given to the need for the economic impacts to be minimized, rather than other aspects, i.e., environmental and social impacts, which are very important but beyond the scope of this study. From a flood impact perspective, capturing and quantifying indirect and intangible effects, i.e., disruption to businesses, communities and loss of life, are not easily tracked and often unobtainable. However, it can potentially be significant and has to be discussed briefly as it may cause an underestimation of flood impact assessment. Therefore, indirect effects, losses, and their costs will then be calculated as a percentage of direct flood damage, as proposed by Lekuthai and Vongvisessomjai (2001).

A methodology for flood management will be found on mathematical modelling and engineering applications, in an integrated way rather than fragmented ones. Conducting a flood simulation and its constituent parts, both models, namely SWAT and 1D/2D SOBEK, will be used to simulate two storm events, i.e. 2001 event and extreme event (1% annual probability of exceedance), and integrated through a graphical user interface intended for use by non-specialists (Kuntiyawichai et al., 2010b). However, issues on uncertainty analysis, although it always links to flood modelling, are not examined as their complexity places them beyond the scope of this study. The package of flood modelling has been developed as a tool to evaluate the effects of hydrologic and hydraulic components on flood generation throughout the Chi River Basin. The results are only used to inform and do not attempt to provide coverage of prevailing policy and legislative settings.

The results shown in this study will present a macro picture of likely flooding including its impact, and give confidence that active flood mitigation is possible once the proposed alternatives are implemented and taken as possible final results in case of well-targeted investments in flood management (Kuntiyawichai et al., 2010a). However, specific findings cannot always be generalised and may not be transferable, i.e., a very cost-effective mitigation measure does not necessarily imply that it will be cost-effective in other river basins.

1.3.3 Hypotheses

One of the first steps towards a thorough explanation of the observed reality is the formulation of hypotheses. Whereas no hypothesis is regarded as true until it is scientifically tested and proven for its validity, and each of which is plausible to a given degree. To come up with a solution to the often repeated flood problem statements, the following hypotheses are formulated and put forward for the study.

- to approach the question of how the flood situation across the Chi River Basin can be alleviated, it is hypothesized that comprehensive and integrated flood mitigation measures would have a higher potential to cope with flood damages compared to individual actions. Through greater coherence between different structural measures and reservoir management, the flood affected areas will become more resilient to man-induced hydrological changes;
- in the context of the impact of flood hazards and risks, there is a contemporary debate to which its disastrous consequences are not solely due to natural phenomena, but rather result from anthropogenic effects. In light of this, it is in the form of anthropogenic land use changes, which affect the spatial and temporal occurrence of water-related extreme events;
- consistent with the aforementioned hypotheses, both proper runoff conveyance and superior land management practices would increase peak flow attenuation effects and consequently reduce downstream flood risk;
- although the optimal set of mitigation measures might require an enormous amount of initial investment, however, the longer term costs in terms of damage associated with extreme flood threat will be much lower;
- above all, one cannot overlook the importance and necessity of an integrated flood modelling approach, which can be distinguished into two ramifications, for investigating specific problems to localised areas. In case of flood inundation simulation, the 1D/2D SOBEK model can perform well and might provide suitable alternative solutions. Whilst the SWAT model can lead to guidance in hydrological assessment of the processes of land use changes.

To determine whether the hypotheses are supported, the principal objective and a series of specific goals would have to be accomplished as described in the next section.

1.3.4 Research objectives

Quite often, vigorous efforts designed to reduce the effects of floods with their narrow focus on flood mitigation benefits, lead to adverse impacts elsewhere. Therefore, flood management in the broadest context integrating both vulnerability reduction and resilience intensification with provision of adequate, technically, and economically sound measures, is required.

In relation to the research questions, the overarching objective of this research is to develop an integrated modelling framework with certain tools and techniques for flood management in light of land use and its changes in the Chi River Basin. In accordance with what is stated above, the more specific objectives of the research are to:

- analyse the nature of the local problems of chronic flooding at a river basin-wide scale;
- identify the most vulnerable areas of flooding to target areas for optimal protection to inhabitants, public and private property;
- understand shortcoming of the existing flood management practices in the Chi River Basin;
- identify the extent to which flood mitigation measures have to be carried out and indicate how these may be assessed in more detail;
- determine design floods for the possible chance of occurrence as a function of the desired level of protection in combination with the implementation of a range of selected flood and land use management measures;

- develop a modelling framework to be able to simulate the effect of land use and land use changes on floods in the Chi River Basin using scenarios by combining a water balance model and a flood model;
- evaluate the various options to improve flood management and to achieve an 'optimal' flood management package to encounter the recurrent problems of flooding in the local conditions of the Chi River Basin;
- evaluate the effects of land use change and flood management measures on the extent of flooding and damages.

1.3.5 Methodology

With chronic flood disasters, who is to blame? Although it would be so easy to blame the human interventions and activities in the river basin and, in turn, for the flood victims to blame themselves, the threat still exists and will occur again and again in the area unless some scheme of mitigation is implemented. Therefore, it is time to introspect their own involvement in the unfolding flood devastation in order to spark a sense of hope critical rather than a sense of doubt and despair. In an attempt to take actions that prevent damage from occurring, a series of cascading steps that interconnect with each other and cover technical and spatial measures is attempted here:

- *calibration highlights.* The study included the development of a calibrated hydrological and hydraulic model, which is based on an actual flood event in September 2001, representing the characteristics of the Chi River Basin;
- *design rainfall event.* A design rainfall has been used as input to estimate the design flood, and derived by fitting extreme value distribution;
- *design flood determination.* A design flood is defined by its probability of exceedance. In this study, the exceedance frequency (safety standard) is fixed at 1% annual probability of exceedance for the potentially affected people;
- *coupled hydrologic and hydraulic models.* The water balance of sub-basins is simulated in daily time steps and its output then serves as input for flood simulation with half-hour time step;
- *technical assessment for flood management.* Assessment of hydrologic and hydraulic behaviour of the river basin is essential to enhance confidence in identifying flood management options, i.e. confidence that an unforeseen circumstance would not make the option ineffective;
- *assessment of flood damage costs.* After assessing the future effectiveness of any specific flood mitigation option, possible reduction of direct flood damage from the full range of possible floods would then have to be evaluated;
- *detailed assessment of preferred options.* As a result of the identification of the benefits that mitigation options deliver elsewhere in the river basin, i.e. their technical viability and economic effectiveness, this step is used to prioritise the options, which enables the selection of a set of technically robust and cost effective solutions that could be optimised;
- *anticipated potential changes in land use on flood hydrology.* A provision of long-term land use change predictions is considered desirable to investigate whether the effect of land use change on flood regimes is feasible within the safety margins of the optimal set of mitigation measures.

1.4 Structure of the thesis

To achieve the desired complex question answering, this thesis consists of eight chapters to provide clearly structured guidance to ensure that an optimal solution of flood management is put in place in a balanced manner. The following paragraphs outline the content of the chapters.

Chapter 1 gives a general overview of the research, which includes concise explanations of its relevance, the questions that are addressed, scope and preliminary guidelines, the hypothesis, research objectives and methodology.

Chapter 2 provides a brief country profile of weather and climate, and the status of land and water resources development. It also highlights the main physical aspects of the Chi River Basin. This chapter ends with specific information about past, current, and foreseeable future development activities in the study area, which can serve as a basis for actions to avoid improper incidents.

Chapter 3 briefly explains general land and water-related details of the Chi River Basin, including an overview of climate and hydrology. The chapter also addresses a description of water management, water use, and supply characteristics, as well as the coordinated use of land and different water use practices. A summary of current ways of managing flood and irrigation system delivery abilities is also elaborated in this discussion, as well as their institutional arrangements.

Chapter 4 explores the literature that is relevant to hydrological impacts of land use change, together with some remarks to further follow their modelling ambitions.

Chapter 5 proceeds with the mathematical modelling, including the descriptions and detailed breakdowns for model selection. The section also comprises a combination of hydrologic and hydraulic simulation models through the connection by Delft-FEWS, and furthermore describes the configuration of the module inside Delft-FEWS. Moreover, this chapter also provides an overview of GIS and its application for addressing land and water development.

Chapter 6 underlines how models help to describe a real application scenario which ultimately leads to an integrated and optimised solution to the provision of flood management measures for the Chi River Basin, and addresses issues related to its evaluation. Part of this chapter is also dedicated to model calibration and validation. In addition, it also includes an overview of the procedures and applications of the key principles involved in statistical analysis in hydrology to further characterize extreme rainfall events, based on their probability of exceedance.

Chapter 7 is devoted to consideration of anticipation when executing the optimal solution of flood management. It also examines the potential consequences of man-made alterations on the hydrological regime, e.g. through land use change.

Chapter 8 wraps up and draws conclusions on knowledge gained from the study. Thereafter, recommendations for future research priorities/directions are given.

2 Background

2.1 General overview of Thailand

2.1.1 Introduction

Thailand is a humid tropical country in Southeast Asia in the centre of the Indochina Peninsula between latitudes 5°27' and 20°27' N and longitudes 97°22' and 105°37' E. The boundaries are shared with the Gulf of Thailand and Malaysia in the Peninsular South, Myanmar (Burma) to the West and Northwest, Laos to the Northeast, and Cambodia to the Southeast (Figure 2.1). The total land area is approximately 51.3 million ha (http://www.bot.or.th/English/EconomicConditions/Thai/genecon/Pages/Thailand_Glance.aspx) with a population of 67.3 million inhabitants (September 2010 estimate) (http://web.nso.go.th/index.htm). The length of Thailand from North to South is about 1,620 km and the widest part from East to West is about 775 km.

Figure 2.1 Map of Thailand (http://www.ldd.go.th/FAO/z_th/thmp111.htm)

Thailand is divided into four regions: the North, the Northeast or the Khorat Plateau, the Central Plain or the Chao Phraya Basin, and the South or the southern Isthmus. The North of the country is a mountainous region characterized by natural forest, ridges and deep narrow, alluvial valleys. The Northeast consists of the Khorat Plateau, bordered to the East by the Mekong River, it is a semi-arid region characterized by rolling surface and undulating hills. It comprises several small river basins, which drain into the two principal rivers Chi and Mun, and through these rivers flow into the Mekong River. The centre of the country is dominated by the predominantly flat Chao Phraya River valley, which runs into the Gulf of Thailand; it is a lush, fertile valley. It is the richest and most extensive rice-producing area of the country and is often called the 'Rice Bowl of Asia'. Bangkok, the capital of Thailand, is located in this region. The South consists of the narrow Kra Isthmus that widens into the Malay Peninsula; it is hilly to mountainous, with thick virgin forests and rich deposits of minerals and ores. This region is the centre for the production of rubber and cultivation of other tropical crops.

2.1.2 General climatic conditions

The climate of Thailand is tropical and therefore warm throughout the whole year. At the same time it is dominated by the monsoon winds that bring about the seasonal changes from wet to dry, creating three seasons in the North and the central areas and two in the South. It can be distinguished into three distinct seasons:

- *rainy or southwest monsoon season (mid-May to mid-October)*. The Southwest monsoon prevails over Thailand and abundant rainfall occurs over the country. The wettest period of the year is August to September. The exception is found in the Southern Thailand east coast where abundant rainfall remains until the end of the year that is the beginning period of the Northeast monsoon and November is the wettest month;
- *winter or northeast monsoon season (mid-October to mid-February)*. This is the mild period of the year with quite cold weather in December and January in upper Thailand but there is a large amount of rainfall in Southern Thailand east coast, especially during October to November;
- *summer or pre-monsoon season, mid-February to mid-May*. This is the transitional period from the Northeast to Southwest monsoons. The weather becomes warmer, especially in upper Thailand and April is the hottest month.

The two major atmospheric circulations affecting Thailand are the Northeast monsoon, with winds from northerly and easterly directions, and the Southwest monsoon, with winds from Southern and Western directions. The onset of monsoons varies to some extent. Southwest monsoon usually starts in mid-May and ends in mid-October while the Northeast monsoon normally starts in mid-October and ends in mid-February. The Southwest monsoon brings humid air from the South China Sea to the East Coast of Peninsular Thailand, where heavy rainfall occurs until early January, especially at the windward side of the mountains. It moves northwards rapidly and lies across the southern part of China around June to early July, that is the reason of dry spells over upper Thailand. The Northeast monsoon brings the cold and dry air from the Siberian anticyclone over major parts of Thailand, especially over the Northern and North-eastern parts. In the South this monsoon causes mild weather and abundant rainfall along the eastern coastline (Figure 2.1).

Not only monsoon but also tropical cyclones cause rainfall in Thailand. Tropical cyclones, which often occur in the southern part of China Sea between September and October, predominantly move towards the Northeast, the East and the central area of Thailand. In this case, heavy rainfall and long continuous periods of rainfall occur in a relatively large area due to wind and clouds disturbance.

2.1.3 Surface temperature

Upper Thailand i.e. the Northern, North-eastern, Central and Eastern parts usually experience a long period of warm weather because of the inland location and tropical latitude zone. From March to May, the hottest period of the year, maximum temperatures usually reach 40 °C or higher except along coastal areas where sea breezes will moderate afternoon temperatures. The onset of the rainy season also significantly reduces the temperatures from mid-May and they are usually lower than 40 °C. In winter the outbreaks of cold air from China occasionally reduce temperatures to fairly low values, especially in the northern and north-eastern parts where temperatures may decrease to near or below zero. In the southern part temperatures are generally mild throughout the year because of the maritime characteristic of this region (Table 2.1). The high temperatures, common to upper Thailand occur seldom. The daily and seasonal variations of temperatures are significantly less than those in upper Thailand.

Table 2.1 Mean seasonal temperatures (°C) in various parts of Thailand for the period 1971 to 2000

Region	Winter season (October - February)	Summer season (February - May)	Rainy season (May - October)
North	23.1	28.0	27.3
Northeast	23.9	28.5	27.7
Central	26.1	29.6	28.3
South	26.6	28.2	27.6

Source: http://www.tmd.go.th/climate/climate_03.html

2.1.4 Rainfall

Upper Thailand usually experiences dry weather in winter (October - February) because of the Northeast monsoon, which is a main factor that controls the climate of this region. The later period, summer (February - May), is characterized by gradually increasing rainfall with thunderstorms. The onset of the Southwest monsoon leads to intensive rainfall from mid-May until early October. Rainfall peak is in August/September, when some areas will be flooded.

The rainy season in the Southern part is different from upper Thailand. Abundant rain occurs during both the Southwest and Northeast monsoon periods. During the Southwest monsoon the Southern Thailand West Coast receives much rainfall and reaches its peak in September. On the contrary, much rainfall in the Southern Thailand East Coast with its peak in November remains until January of the following year, which is the beginning of the Northeast monsoon (Table 2.2).

According to a general annual rainfall pattern, most areas of the country receive 1,200 - 1,600 mm/year. Some areas on the windward side, particularly Trat Province in the eastern part and Ranong Province in the Southern Thailand West Coast have more than 4,000 mm/year. Annual rainfall less than 1,200 mm/year occurs in the leeward side areas, which are clearly seen in the central valleys and the uppermost portion of the southern part.

Table 2.2 Seasonal rainfall (mm/season) in various parts of Thailand for the period 1971 to 2000

Region	Season			Annual average	Annual rainy days
	Winter	Summer	Rainy		
North	106	183	952	1,240	123
Northeast	72	214	1,090	1,370	117
Central	124	187	903	1,220	113
South					
- East coast	759	250	707	1,720	148
- West coast	446	384	1,900	2,730	176

Source: http://www.tmd.go.th/climate/climate_04.html

2.1.5 Relative humidity

Thailand is covered by warm and moist air in most periods of the year except for the areas further inland the relative humidity may significantly reduces in winter and summer (Table 2.3). For example, the extreme minimum relative humidity values showed only 9% at Suphanburi and Chiang Mai on 25 February 1972 and 24 March 1960, respectively. In the southern part, which is maritime the humidity is higher.

Table 2.3 Relative humidity (%) in various parts of Thailand for the period 1971 to 2000

Region	Season			Annual average
	Winter	Summer	Rainy	
North	74	64	81	75
Northeast	69	66	80	73
Central	70	69	79	75
South				
- East coast	80	77	79	79
- West coast	78	76	84	80

Source: http://www.tmd.go.th/climate/climate_05.html

2.1.6 Land and water resources

In the context of resources such as land and water, there is a growing concern due to the patterns of over-exploitation and broader anthropogenic change. To keep track of the current situation, an overview of land and water resources in Thailand is summarized below.

Land resources

A number of land resource inventory surveys have been undertaken, primarily by the Land Development Department under the Ministry of Agriculture and Cooperatives, over all parts of Thailand. It has been widely applied in Thailand to identify opportunities and constraints to land use and is contained in a GIS data base. The land resource inventory comprises two sets of data:

- an inventory of six land use patterns, as shown in Table 2.4. The majority of land is used for agriculture, i.e. paddy fields, cash crops, vegetables, fruit crops and trees. It covers approximately 23.5 million ha or 45.8% of total land. In order to better understand the economic significance of agriculture within the entire area, the agricultural area in each region may be divided into five distinctive regions as listed in Table 2.5;

Table 2.4 Land resources in Thailand

Land use pattern	Area in 1,000 ha	Area in %
Paddy field	13,400	26.1
Cash crop	7,750	15.1
Vegetable	14	0.0
Fruit crops and trees	2,340	4.6
Forest	19,700	38.5
Miscellaneous (i.e. community, reservoir, bare land, etc.)	8,060	15.7
Total	51,300	100.0

Source: http://www.ldd.go.th/EFiles_html/land%20resource/ed0200.htm

Table 2.5 Agricultural areas by region in Thailand

Agricultural area in each region	Area in 1,000 ha	Area in % of agricultural land
Northeast	9,640	41.0
North	5,340	22.7
Central and West	3,740	15.9
South	2,910	12.4
East	1,880	8.0
Total	23,500	100.0

Source: http://www.ldd.go.th/EFiles_html/land%20resource/ed0200.htm

- a derived assessment identifying the suitability of land for sustained agronomic production (land suitability), from research of soil properties by using the soil map in region level scale 1:500,000. It consists of environmental factors by taking into account its productive potential, physical limitations, including climate, soil conservation needs and management requirements, which can be classified as given in Table 2.6.

Table 2.6 Land suitability for sustained agronomic production in Thailand

Land use pattern	Area in 1,000 ha	Area in %
Suitable for cash crop	10,800	21.1
Suitable for paddy field	13,500	26.3
Suitable for tree in rainfed	2,620	5.1
Unsuitable for economic crop but may be some specific crop	7,970	15.5
Unsuitable for agriculture	16,000	31.1
Wetland	396	0.8
Total	51,300	100.0

Source: http://www.ldd.go.th/EFiles_html/land%20resource/ed0200.htm

Land resources have changed dynamically due to human-induced processes that permanently alter the land. There are four main problems of land resources, which can be summarised as follows:

- misuse of land, i.e. residential and industrial construction on agricultural land, deforestation and encroachment into the river basin conserved area, cultivation of plants that are not suitable to land. The total area of misused land concerns 4.8 million ha;

- land mismanagement is as follows:
    - soil erosion, loss of nutrients, minerals and organic matter. There are approximately 17.1 million ha or 33% of the whole country having this problem;
    - low organic matter. This problem is commonly found covering as high as 30.6 million ha or 59.5% of the total area in the country;
- topology and environment. Examples of this type of problem are as follows:
    - the land is unfertile resulting from its salinity. Moreover, it is flooded in the rainy season but drought covers the whole area in the dry season;
    - the problem is mainly contributed from the characteristic of soil itself as well as the natural environment of that area;
    - the area is severely flooded all year round and contains too much organic matter;
    - the area is constituted mainly with gravel soil, fertility is too low while its structure is not suitable for cultivation;
- problem soils. Natural condition of soil:
    - *acid sulphate soils*. In the central lowland region alone there are 7 provinces facing this problem. The area of severely acid to moderately acid soil is 368,000 ha with the average yield of rice at 1.4 tons/ha;
    - *saline soils*. In the north-eastern part of Thailand, this type of soil is scattered all over. The total area ranging from severe, moderate, to low salinity amounts 2.9 million ha or 16.7% of the north-eastern land area;
    - *saline and acid sulphate soils*. In the South, there are as much as 4 million ha of saline soil and 160,000 ha of acid soil. Thirty percent of the saline soil and almost all acid soils are either used as paddy field or left unused. The average yield from this kind of soil amounts only 0.9 - 1.9 tons/ha.

Moreover, the problem soils can also be categorised into different types of specific problem soils as shown in Table 2.7.

Table 2.7 Different types of specific problem soils

Problem Soils	Area in 1,000 ha	Area in %
Salt affected soil:		
- Coastal saline soils		
• Potentially acid	462	1.6
• Non-potentially acid	116	0.4
- Inland saline/Sodic soils		
• Extreme saline soil	283	1.0
• Moderate saline soil	590	2.0
• Low saline soil	2,023	6.9
Sandy soil:		
- Extreme sandy soil, no organic stratum	1,058	3.6
- Extream sandy soil with organic stratum	82	0.3
Acid sulphate soil	852	2.9
Organic soil	81	0.3
Shallow soil		
- Laterrite soil and conglomerate soil	5,087	17.5
- Soil with stone	2,772	9.5
- Soil with calcium bi carbonate	347	1.2
Slope complex	15,385	52.8
Total	29,140	100.0

Source: http://www.ldd.go.th/EFiles_html/land%20resource/ed0300.htm

Water resources

In the past three decades, the continuous rapid economic development in Thailand stimulated an explosive increase in the demand for water services, such as for power, irrigation, domestic, and industrial water supply. The Royal Thai Government has devoted important resources to serve these demands. Water management approach emerged with emphasis on expansion of access to services for those mentioned purposes. This approach was successful in giving millions of Thai access to potable drinking water, water to produce cheap and abundant food, and to generate hydropower. However, this approach is no longer appropriate because water has become increasingly scarce. The Government now faces a different and more complex set of challenges, comprising both supply and demand-side questions (Sethaputra et al., 2001):

- is the resource base, including both water and river basins, being managed in a sustainable manner?
- are there opportunities for more effective management of existing sources of supply?
- who will be allocated the water, and how will it be allocated?
- who will provide and deliver services, and who will pay for them?

Recently, Thailand's water resources, such as surface water and groundwater, have encountered three critical issues: drought or water shortage, floods, and water quality problems. All these problems have caused severe damage and negative impact on the economy and people's way of life. Thailand is divided into 25 river basins and 254 sub-basins, with total coverage area of 51.4 million ha (Figure 2.2 and Table 2.8) (Department of Water Resources, 2005b). In relation to water usage, the major water requirements include domestic consumption, agriculture, and industry. Thailand's water resources in detail include the following.

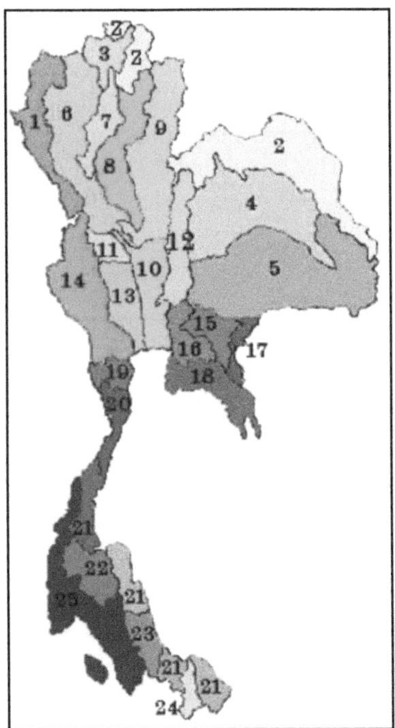

Figure 2.2 Twenty five major river basins of Thailand (Royal Irrigation Department and Department of Water Resources, 2008)

Table 2.8 Twenty five major river basins of Thailand

River basin no.	River basin name	River basin area in 1,000 ha
01	Salwin river basin	1,800
02	Mekong river basin	5,740
03	Kok river basin	790
04	Chi River Basin	4,950
05	Mun river basin	6,970
06	Ping river basin	3,390
07	Wang river basin	1,080
08	Yom river basin	2,360
09	Nan river basin	3,430
10	Chao Phraya river basin	2,010
11	Sakae Krang river basin	519
12	Pa Sak river basin	1,630
13	Tha Chin river basin	1,370
14	Mae Klong river basin	3,080
15	Prachin Buri river basin	1,050
16	Bang Pakong river basin	798
17	Tonle Sap river basin	415
18	East Coast - Gulf river basin	1,380
19	Phetchaburi river basin	560
20	Prachupkiri - Khan Coast river basin	675
21	South East Coast river basins	2,640
22	Tapi river basin	1,220
23	Thale Sap Songkhla river basin	850
24	Pattani river basin	385
25	South West Coast river basin	2,120

Source: Office of the National Water Resources Committee (2000)

Surface water

The total annual volume of water from rainfall in all river basins is approximately 800 billion m^3, of which 75% or around 600 billion m^3 is lost through the processes of evaporation and infiltration. The remaining 25% or 200 billion m^3 flows directly as surface runoff into rivers and streams (Suiadee, 2002). Table 2.9 shows the data on surface water resources in Thailand.

Table 2.9 Surface water resources in Thailand

Regions in Thailand	River basin area in 1,000 ha	Average annual rainfall in mm/year	Amount of rainfall in billion m^3	Amount of runoff in billion m^3
Northern	17,000	1,280	217	65
Central	3,000	1,270	38	8
North-eastern	16,900	1,460	247	37
Eastern	3,400	2,140	73	22
Western	4,000	1,520	61	18
Southern	7,000	2,340	164	49
Total	51,300		800	199

Source: http://www.icid.org/v_thailand.pdf

Groundwater

Groundwater is an important resource for water supply in Thailand. It is mainly recharged by rainfall and seepage from rivers. In addition, it is widely used for urban and rural domestic water supply and also for agriculture and industrial purposes. Public water supplies for one-fifth of the nation's 220 towns and cities and for half of the 700 municipalities are derived from groundwater. Previous water balance studies of the river basins under favourable geologic conditions such as those in the Northern Highlands, the Upper Central Plain and along the Gulf Coastal Plain had estimated that 12.5 - 18.0% of rainfall infiltrates the soils and about 9% of rainfall reaches aquifers (Suiadee, 2002).

The quantity and quality of groundwater vary according to local hydrological conditions. Usually large and high yielding aquifers occur in alluvium and terrace deposits. Moreover, groundwater also exists within limestone formations, sand stones and some types of shales. In river basins under unfavourable geologic conditions such as in the Lower Central Plain where Bangkok is situated, about half of the area is covered by thick marine clay and in the Khorat Plateau where its central part is covered by impervious shale. It is estimated that here only 5 - 6% of rainfall reaches aquifers.

The first systematic government programme for groundwater investigation and development began in 1955 in the north-eastern region where water shortage is critical for 6 - 8 months per year. The programme objectives were to provide potable groundwater for rural water supply and to evaluate essential information required for proper development of the groundwater sources. Similar programmes were later conducted in other regions throughout the country. The Department of Mineral Resources has long been involved in groundwater investigation. The Department has conducted studies and was involved in groundwater development. Systematic investigations leading to the aquifer system analysis have been initiated recently, except for the Bangkok Metropolitan Area where a program of detailed groundwater investigation and simulation was implemented together with the monitoring of groundwater levels and land subsidence (Suiadee, 2002). As a result, it has been found that the largest sources of groundwater in Thailand are found in the Lower Central Plain, especially in Bangkok and surrounding provinces.

Starting from 1982 up to 2002, more than 200,000 groundwater wells were installed by both the government and the private sector with total capacity of about 7.6 million m^3/day (2.7 billion m^3/year). It is estimated that 75% of domestic water is obtained from groundwater sources, which serve approximately 35 million people living in villages and urban areas (Pattanee, 2006).

2.1.7 Land and water development

Land and water are two valuable and essential resources, related to the multiple aspects of human life. Therefore, it is necessary to pay extra attention to land and water development by considering not just only environmental but also socio-economic view points. Land and water development are major challenges for the socio-economic development, particularly in areas where water-related hazards affect people most, for the improvement of their living conditions.

The concept of land and water development becomes important in Thailand. It is increasingly being deployed in various development programmes to manage the land and water resources, for example soil and water conservation, rainfed dryland agriculture, land reclamation, control of shifting cultivation, vegetation cover improvement, etc.

However, fragmentation, overlapping among various water-related agencies and the lack of appreciation of the impact of water development on land are major causes of ineffective management. The environmental impacts of land and water development have not been taken as integral parts of such development. Hence, more integrated and coordinated plans would be required for the development of an approach to the specifics of land and water development as well as the broader aspects of agriculture planning and management.

Issues related to water resources development and management in Thailand encompass a number of different actions required to manage and control water resources. Such actions involve flow stabilization for irrigation, flood control, hydropower, domestic, industry, and various miscellaneous uses. However, in view of an agricultural based country, development of water resources has primarily been a significant factor to support agriculture and related activities over the past decades, and subsequently expanding through the manufacturing and service sectors in recent decades. Moreover, population growth and urbanization, associated with changes in production and consumption, have also placed exceptional demands on water resources. Meanwhile, water use became careless and wasteful since the price of water is relatively low for people compared to many other necessities. Taking all these facts into consideration, it is true that water-related problems (i.e. water conflicts, water pollution, water shortage, floods, and droughts) will become even more complicated (Sethaputra et al., 2001).

In essence, the current view of the water problem issue is not only limited to inadequate provision, but also includes inefficient allocation and low efficiency. In an attempt to tackle such problems, an array of governance and management measures would have to be more integrated with a significant increase in the level of public participation and stakeholder involvement, including private sector, relevant Non-Governmental Organizations (NGO), Community Based Organizations (CBO), and other relevant stakeholders. However, the current water resources management issues and challenges in Thailand are still related to proper management and allocation of water resources, partnership actions and integrated approach, and sustainable use and development of water resources.

Moreover, the Royal Thai Government set up the policies on water resource management in order to provide solutions for development or upgrading. Those policies will be verified and presented into three main categories based on area functions, i.e. the upper (forest area), middle (agricultural area and community), and lower (downstream including coastal area) river basin. An update information concerning all natural resources and other related matters including environmental aspects shall be examined, monitored and technically verified for data base establishment, to generate management and other plans in the most effective and proper processes from the upper through lower downstream areas. In regard to the institutional arrangements, the Ministry of Natural Resources and Environment is responsible for the policy and overall planning of natural resources including water resources, while the Ministry of Agriculture and Cooperatives is mainly responsible for the implementation and operation of the infrastructure for the agricultural areas (Ministry of Natural Resources and Environment, 2008).

2.2 Brief description of the study area

2.2.1 Chi River Basin

The Chi River Basin, situated in the northeast of Thailand at latitudes 15°30' and 17°30' N, longitudes 101°30' and 104°30' E, is a semi-arid area characterized by a rolling surface and undulating hills. The area covers 4.9 million ha with a population of 6.6 million. The Chi River is the longest river in Thailand, which is 946 km long. However, it carries less water than the Mun River, which is the second longest river. The river rises in the Phetchabun Mountains, then runs East through the provinces of Chaiyaphum, Khon Kaen and Maha Sarakham, before turning South in Roi Et and running through Yasothon to meet the Mun River in Kanthararom District of Sisaket Province at about 100 km upstream of the confluence with the Mekong River (Figure 2.3).

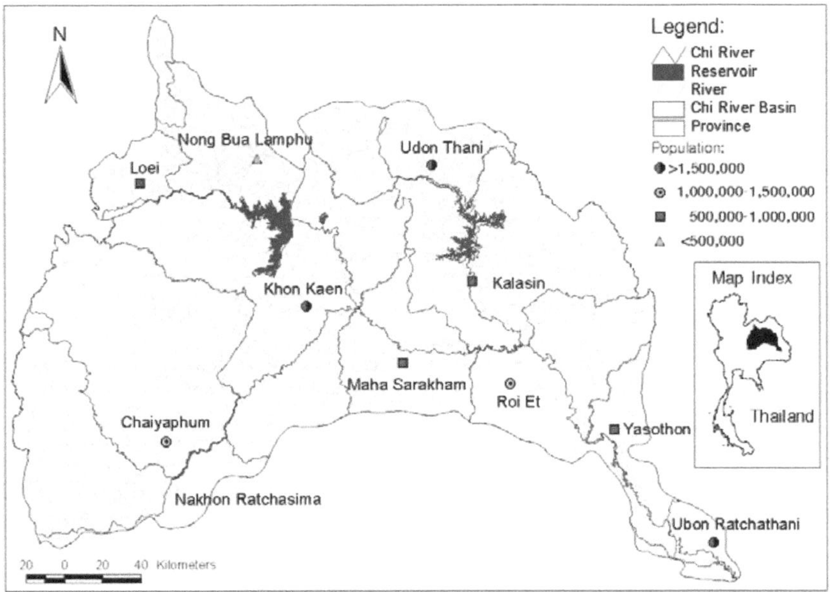

Figure 2.3 Chi River system (Regional Centre for Geo - Informatics and Space Technology, 2001)

2.2.2 Characteristics of the study area

Topography

The Chi River Basin originates from the mountainous part, which is the Phetchabun mountain range in the west of the river basin with an average elevation of about 1,300 m+MSL (Mean Sea Level). The majority of the area in Chi River Basin consists of plains with elevations generally of 200 m+MSL at the lower part, especially the downstream part.

The slope of Chi River Basin is steep at the upstream mountain area and is flat at the lower part, especially near the confluence with the Mun River (Figure 2.4). Relief is mainly rolling, and rivers are generally incised several metres below the surface of the mainly sandstone plateau. Most of the land is cultivated for rice in the lowland part, while the field crops, i.e. cassava, sugarcane, kenaf, maize, etc. are widely found in the uplands.

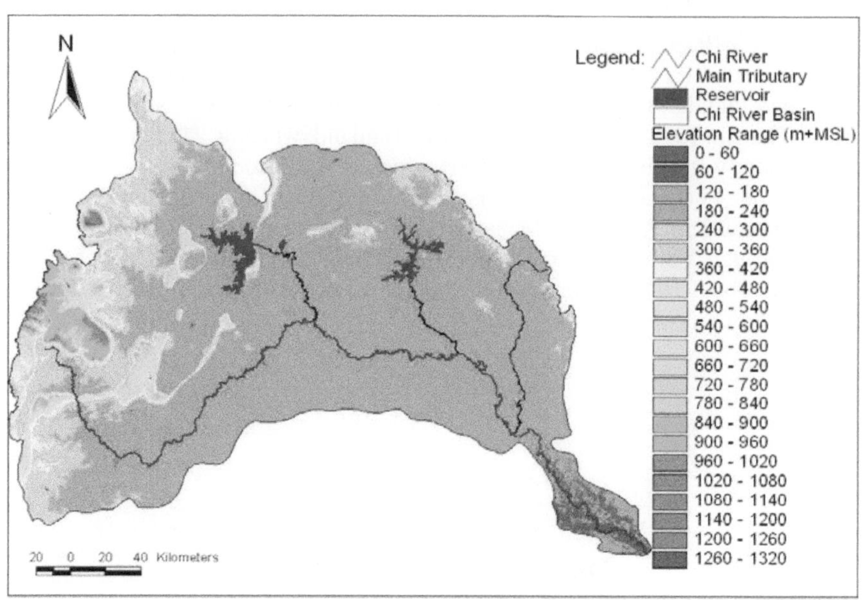

Figure 2.4 Topographic map of Chi River Basin

Soils and geology

The soil throughout the Chi River Basin is sandy loam with poor drainage (Figure 2.5), which makes most of it not suitable for rice cultivation.

Soil drainage classes are assigned in the Chi River Basin in order to categorize the soil within each soil mapping unit on the basis of field observations during the soil survey as shown in Figure 2.6. The properties of soil drainage classes can be summarized as follows:

- excessively drained soils, which are almost exclusively coarse-grained soils, with very high permeability;
- well drained soils, which are generally coarse grained or shallow over bedrock, and are mostly on steep slopes;
- poorly drained soils, which are generally wet due to several factors such as fine texture, flat slope, and high regional groundwater table.

In view of geology, the Chi River Basin is situated in the Khorat Plateau. An extensive outcrop of Mesozoic rocks called the Khorat Group occurs on this plateau. A large part of the Khorat Plateau is blanketed by consolidated sediments, which consist of sand stone, shale, silt stone, conglomerate and rock salts. These rocks, in turn, are overlain by a sequence of alluvial to terrace unconsolidated sediments that cover much along the floodplain in the Chi River Basin. The geological description is shown in Figure 2.7 and Table 2.10.

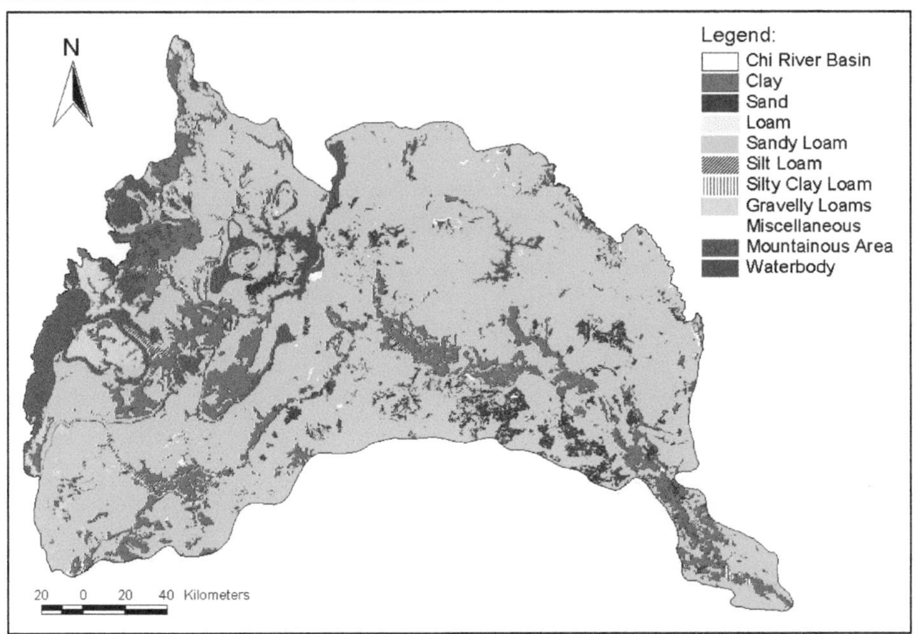

Figure 2.5 Soil map in Chi River Basin (Regional Centre for Geo - Informatics and Space Technology, 2001)

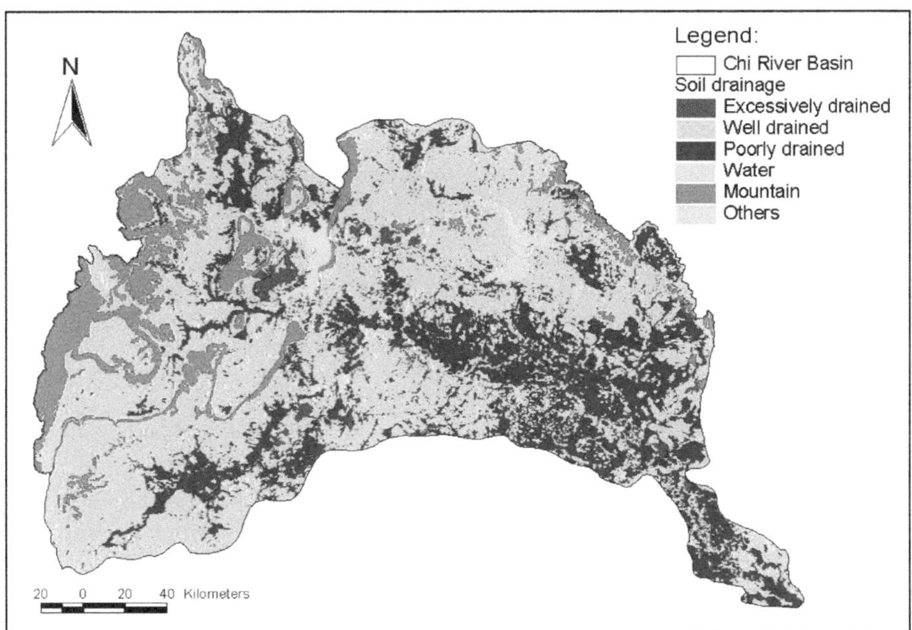

Figure 2.6 Soil drainage in Chi River Basin (Regional Centre for Geo - Informatics and Space Technology, 2001)

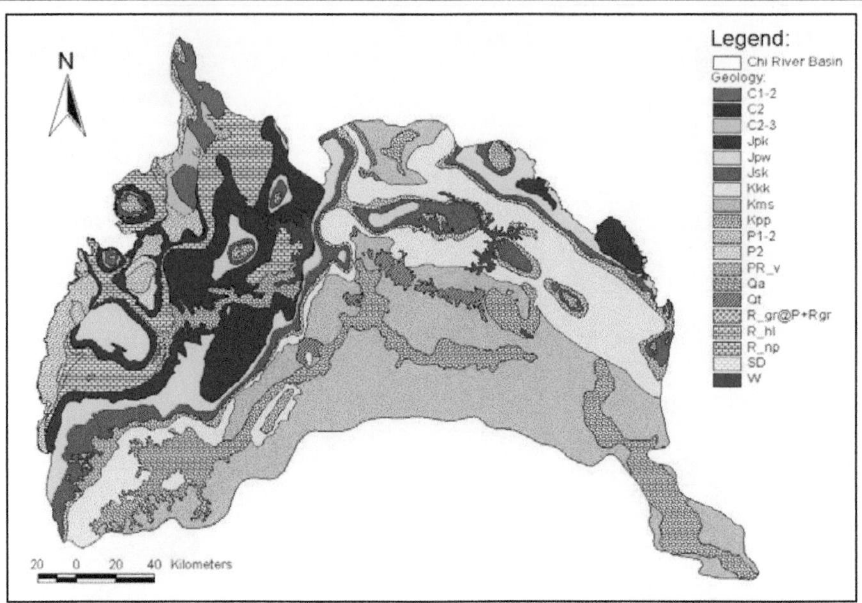

Figure 2.7 Geological map of Chi River Basin, for explanation of the legend see Table 2.10 (Regional Centre for Geo - Informatics and Space Technology, 2001)

Table 2.10 Geological description of Chi River Basin (Regional Centre for Geo - Informatics and Space Technology, 2001)

Symbol	Geological description
C1-2	Sandstone, greywacke, and shale
C2	Shale, sandstone, conglomerate, quartz conglomerate, and quartz sandstone
C2-3	Sandstone, shale, chert, greywacke, and conglomerate
Jpk	Purplish-red siltstone, fine grained sandstone, shale, and conglomerate
Jpw	White to light brown quartz sandstone; siltstone, and shale
Jsk	Reddish brown siltstone, mudstone, sandstone, and shale
Kkk	Brown, reddish-brown micaceous sandstone; pale brown micaceous shale, siltstone, and conglomerate
Kms	Brick-red, purplish-red mudstone, siltstone, shale, and sandstone; rock salt with potash, gypsum and anhydrite
Kpp	White, pale orange, yellowish brown, pebble sandstone intercalated with shale and conglomerate
P1-2	Limestone
P2	Bedded to massive limestone, calcareous shale, laminated shale, tuffaceous sandstone, tuff and chert
PR_v	Andesite, rhyolite, tuff, and agglomerate
Qa	Alluvial deposit: river gravel, sand, silt, and clay
Qt	Terrace deposit: gravel, sand, silt, and clay
R_gr@P+Rgr	Granite, granodiorite, and diorite
R_hl	Interbedded shale, mudstone, siltstone, greywacke, argillaceous limestone, basal limestone conglomerate and local volcanic conglomerate of fusulinid bearing limestone quartz, various kind of volcanic rocks, greywacke (the CP), granitoids, mafic plutoni
R_np	Brown to red sandstone, shale, and conglomerate
SD	Quartzite, phyllite, schist, sandstone, shale, and tuff
W	Water

Land use

The Chi River Basin has experienced rapid land use changes, urbanisation increases, and intensive and extensive agricultural land development. As a result, there is an urgent need to preserve the integrity of the river basin through monitoring land use changes and identifying problems. According to this view, overall land use in the Chi River Basin can be divided as presented in Table 2.11 and Figure 2.8.

In fact, the total area of agricultural land is about 3 million ha. Rice is the major crop, which covers approximately 40.7% of the agricultural land. Other crops include 18.8% of field crops, 0.4% of perennial plants, and 0.1% of fruit crops.

Table 2.11 Present land use in the Chi River Basin

Type of land use	Area in 1,000 ha	Area in %
Agricultural land	2,969	60.0
Forest	1,539	31.1
Urban and built-up area	143	2.9
Waterbody	124	2.5
Miscellaneous area	173	3.5
Total	4,948	100.0

Source: Department of Water Resources (2006)

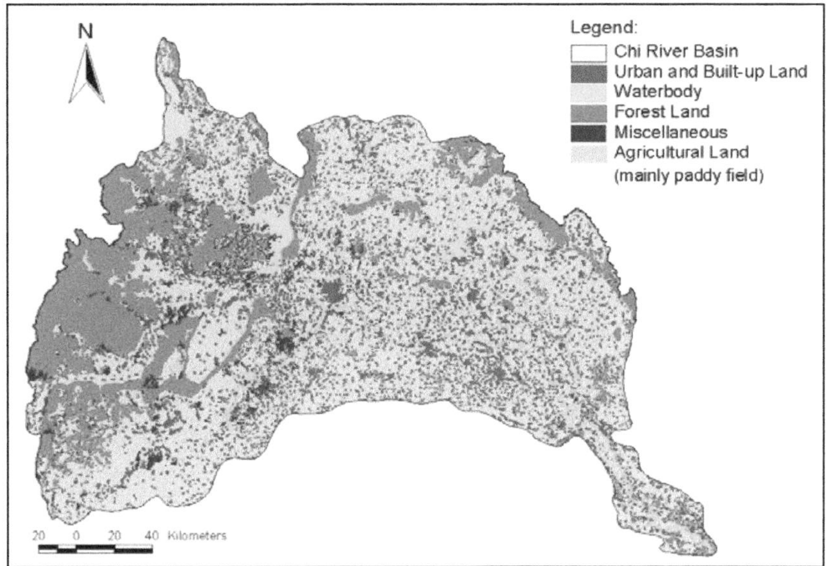

Figure 2.8 Present land use in Chi River Basin (Regional Centre for Geo - Informatics and Space Technology, 2001)

Floods and droughts in general

The Chi River Basin has always been subject to flooding, and it became increasingly severe in the following years: 1978, 1980, 1995, 2000, 2001, 2002, 2003, 2004, 2005, and 2006 (Royal Irrigation Department, 2008). In recent years (2007 - 2010), the threats of floods still remained, however, the year 2010 flood has caused the most damage in both upstream and downstream parts. High intensity rainfall in the upper part is causing a quick runoff response and results in flash flood, whilst on flat areas downstream it causes stagnant flood. Together with the rising waters of the Mun River that occurs almost at the same time, it has resulted in decreasing in drainage efficiency of the Chi River. Consequently, floodwater is accumulated in the vicinity of Chi-Mun confluence (Figure 2.3 and 2.9), and correspondingly connected to the flood area overflow from Mun River, which is the largest tributary of the Mekong River in terms of the area of the river basin. Table 2.12 illustrates the travel times of flood peaks as they proceed along the Chi River (note: these times are based on experience of previous events).

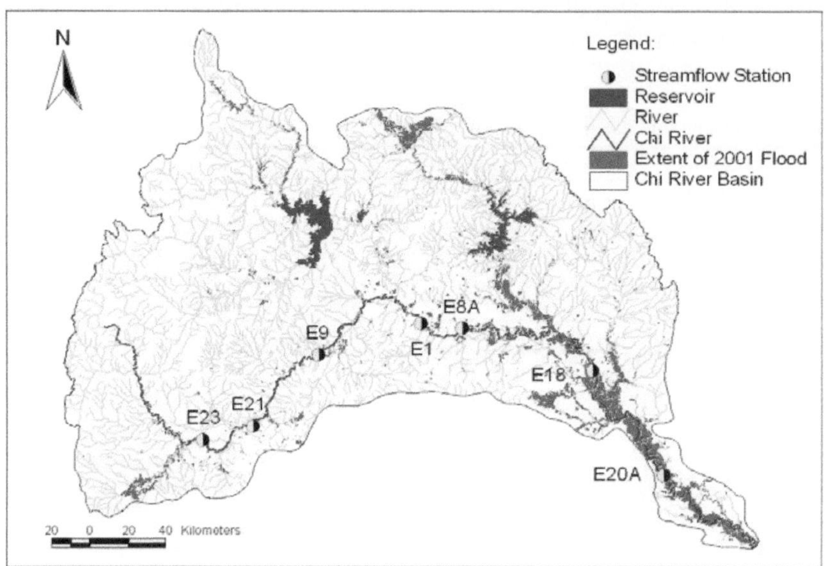

Figure 2.9 Flood prone area in the Chi River Basin, 2001 (Regional Centre for Geo - Informatics and Space Technology, 2001)

Table 2.12 Travel times (hours) of flood peaks in the Chi River

Streamflow station	Distance in km	Year 2000	Year 2001	Year 2002
E23				
↕	60.7	37	24	43
E21				
↕	84.5	72	96	145
E9				
↕	129.4	60	59	59
E1				
↕	30.1	24	42	36
E8A				
↕	30.1	78	117	84
E18				
↕	153.4	84	54	96
E20A				

Source: Royal Irrigation Department (2008)

Drought is also one of the major hydrological hazards in this area. Periodically, severe droughts in this region cause crop losses, reservoir depletion, low flows, and water quality deterioration.

Drought annually occurs in the dry season during March and April. The situation is worse than in the past due to the increase in water demand from various sectors, i.e.

domestic, agriculture, and industry, as consequences of increasing population and economic development. In fact, the cause of drought in the Chi River Basin is deforestation together with soil characteristics and dry climate. Moreover, a side effect of the El Nino phenomenon sometimes worsened the situation by affecting the whole river basin at different levels on shortage of water, especially for those mentioned purposes (Asian Disaster Reduction Center, 1999).

The 2006 statistics show that approximately 77% of the total land area damaged by drought was found in the northeastern part where the Chi River Basin is located, and most of it was under rice cultivation (Department of Disaster Prevention and Mitigation and Department of Agricultural Extension, 2006).

According to the record of drought occurrence and impact collected by the Royal Irrigation Department, out of a total of 8,137 villages in the Chi River Basin, drought damage occurred in 4,808 villages (59.1%). To gain an insight into drought conditions, villages were sorted into two water scarcity categories: village faces water scarcity in agriculture only, and village faces water scarcity in both domestic practices and agriculture. It was found that 2,658 villages (32.7%) were categorised under the agricultural water scarcity category, while 2,150 villages (26.4%) did fall into the latter category (Royal Irrigation Department, 2006).

Population characteristics

In 1993, the total population of the Chi River Basin was 6.2 million. However, there has been an increase in the number of inhabitants, i.e. by 2003 the population was increasing at an annual growth rate of 0.56% to 6.6 million. Overall, about 2.6 million or 39% of the population resides in urban areas, while 61% (4.0 million) lives in rural areas.

The average population density is 134 persons/km^2, but varies greatly from 19 persons/km^2 in the Lam Saphung sub-basin to 212 persons/km^2 in the Lower Part of Lam Nam Phong and Lower Part of Lam Nam Chi sub-basins.

From a population projections point of view, a 30-year projection up to the year 2033 was made. According to that projection, the population of the Chi River Basin will continue to grow at a linear rate. Figure 2.10 illustrates the population projection for the Chi River Basin covering the period of 2013 to 2033 in five-year time steps.

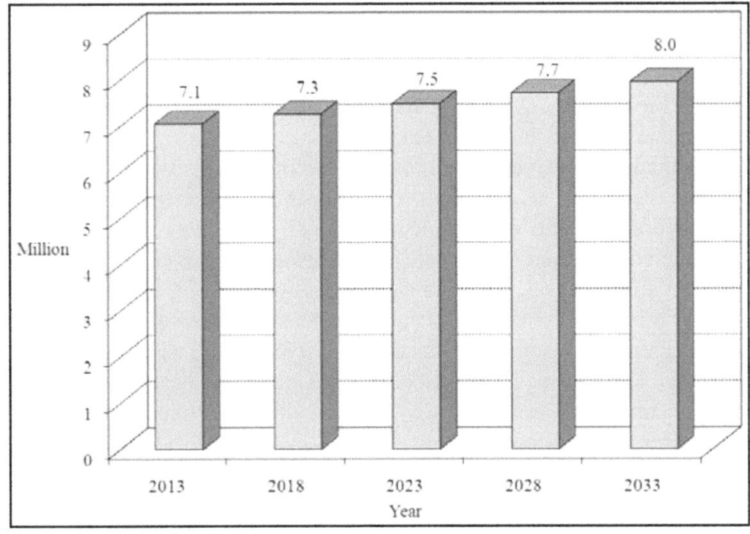

Figure 2.10 Population projection for the Chi River Basin (Department of Water Resources, 2006)

2.3 Developments in the Chi River Basin

To address the issues on water resources management, the long-term consequences of past and present development practices in the river basin have to be analysed. Together with the impact of future development plans, which likely involves an increase in population, social and economic activities, on river basin water dynamics.

2.3.1 Past developments

Without a good historical perspective, water resources management issue might fail to replace the mistakes of the past, since it has an unclear understanding of the dynamics of the complex human-environment system. Therefore, past development trends would have to be emphasized as a lesson to learn as well as an option for the future.

Two thousand years ago, people in the Chi River Basin scatteringly settled in the sparse forests, which were flooded in the rainy season. People enjoyed digging canals and developing reservoirs of water for drinking and agricultural purposes in this abundance period until the 17th century before gradually declined. During that period, they had been displaced by the Angkor civilization in the 10th century, which led to the changes of agricultural techniques. The sparse forest and low-lying areas around the Chi River Basin were drained and reclaimed using a ring dike method. They developed a bucket system for irrigation as well as groundwater reservoirs to hold groundwater for drinking. Settlements were situated around these reservoirs. These presented a significant change of primitive extensive farming pattern into small-lot farming methods centred around individual reservoirs. They grew rice by transplanting seedlings to irrigated paddies as they practice nowadays. As the population gradually increased, numbers of small villages and some large towns were also greater. However, these settlements eventually disappeared due to unknown reasons but they were affected by changing climatic conditions, changing in flooding and drought patterns, and the incursion of the Thai tribes. In the meantime, a lot of cultivated lands eventually returned to forests (Hori, 2000).

The livelihoods of people living in the Chi River Basin highly depend on natural resources such as land, water, and services provided by the surrounding environment. In the past, people had a simple and contented life with different activities such as fishing, and gathering firewood, vegetable, fruit, culinary and medicinal plants from the forests, etc. As stated by Fukui and Hoshikawa (2003), earth bunds were originally built by farmers to store water. They even used turbines to divert water from rivers and ponds for irrigation (Floch et al., 2007). Nevertheless, the situation is changing due to rapid growth in demand for land and water resources.

Similar to many other rivers, population and economic growth is increasing pressure on Chi water resources to meet future demands in domestic, industrial, irrigation, livestock, hydropower, and environmental sectors. Moreover, urban growth has also been one of the contributors to compete for freshwater in this particular area. Although water shortages in most uses are quite well understood among all user groups, but only little was known of the environmental consequences of increased competition for water. Likewise, a decrease in water flows through the river basin has contributed to increases of soil salinity, agrochemical pollution, and biodiversity destruction. Therefore, the changes in the area have consequently affected the opportunities and constraints for resource management at different levels.

In a context where an aggressive agricultural expansion resulted in broad-scale clearing of vast areas, agricultural expansion in the Chi River Basin has involved both extensification where a greater cultivated area utilizing available irrigation facilities and rainfed runoff, and intensification where places greater water demands per unit area (Walsh et al., 1999; Crews-Meyer, 2004). Indirectly, Walsh et al. (2001) revealed that agricultural growth has reduced the forest cover, while chemical fertilizers and pesticides have polluted the waterways, with impacts on hydrological regimes and water quality, respectively.

As a result of the industrial growth, water supply to the industries is also confronted with problems of availability and adequacy. It has placed ever greater demands on Chi River Basin water supply. In part, the demand comes from the direct need for water in industrial processes, and partly from rapid growth in electricity use, which consequently led the government to build hydropower dams on most of the main rivers. With growing water scarcity and increasing demand, dams have therefore tended to serve as a focal point for conflict over the use and management of water. Meanwhile, water quality has become another major issue as a consequence of poorly regulated industrial development of areas in Chi River Basin.

A discussion of past land use trends is also one of the meaningful and equally important aspects of development. In Chi River Basin, land cover and land use have changed extensively in the last decades. Forests have been enormously replaced by agricultural land, and large dams for irrigation and hydropower purposes (Costa-Cabral et al., 2008). Since the forests are cleared, concerns therefore focus on the loss of biodiversity and livelihoods. Besides, it can lead to leaching of nutrients and erosion of sediments, which flow finally into rivers. However, it is still unclear how much land use changes have impacted on the hydrological regime and its extremes (floods and droughts) of the Chi River Basin (Mekong River Commission, 2005).

At last, relevant to progressive development of water resources in the Chi River Basin, the issue of concern has already determined and resulted in implementation of current irrigation systems in order to provide supplementary water, which consists of three types of irrigation schemes: large, medium, and small scales (Intarachai, 2003). Water can be supplied either by gravity or pumps. A number of dams and reservoirs have been built for several purposes such as irrigation, domestic consumption, hydropower, industry, flood control, navigation, fishery, recreation, etc. In the Chi River Basin, there are three dams and reservoirs with a different primary purpose, namely, Ubol Ratana, Lam Pao, and Chulabhorn (more details see Sections 3.1.2, 3.2.2, and 3.4.3). In addition to the three multipurpose reservoirs, several small dams and weirs have also been constructed in this river basin. At present, there are around 1,421 dams and weirs schemes in the Chi River Basin. Approximately 365,200 ha of the total Chi River Basin of 4.9 million ha had been developed as irrigated agricultural area fed by gravity flow (Royal Irrigation Department, 2008). Dams, weirs, and reservoirs have been constructed to induce gravity flow. In addition, dams and weirs are using as water storage reservoirs and flood protection, which is the main cause of change in the hydrological regime of rivers. To manage water resources more efficiently, cooperation and coordination from the concerned sectors is needed, particularly the various government sectors need to work harmoniously. To clarify the various aspects of changes caused downstream when a dam is built on a river, these observations are in agreement with Collier et al. (1996), the modern large dams have as much effect on rivers and human communities as the floods they are intended to control. Consequently, the overall discharge of the Chi River has declined continuously since the start of dam construction. A lot of water is now diverted for irrigation and other purposes while water in reservoirs behind numerous dams provides substantial surface areas for evaporation losses.

2.3.2 Present and future developments

The Chi River Basin has its own unique development on water resources, agriculture, land use, etc. The livelihood of inhabitants in Chi River Basin is mainly sustained through agriculture cultivation, animal husbandry, and trade, all taking place within the relationship that exist between people, animals, and environment.

Formerly, development in a modern sense was thought to be the cause of enormous interruption to the environment, for that reason, very slow progress was made in the fields of material and technological developments in the Chi River Basin. This situation was significantly changed after the occupation by the inhabitants.

In actual fact, industrial and hydropower development in the Chi River Basin are not very expanded and most agriculture is not intensive. Due to a large population, which concerns mainly poor subsistence farmers, there is an obvious need for development and potential for water use conflict, unless development strategies are undertaken with a sound understanding of the full range of alternatives and their consequences for the natural resources of the river basin and the subsistence users who depend on them. The serious disadvantage aspects have been taken into consideration, in the name of development, relevant aspects are briefly described below.

Land use

The Chi River Basin has been facing problems of deterioration in soil fertility due to improper land use practices for decades. Improper land use management and deforestation have resulted in severe erosion in many areas of the river basin. It is found that there are serious soil erosion problems in some areas of non-utilized land and in the upstream deforested hills. Moreover, saline and acid soils in several areas also need special treatment for successful utilization.

A number of relevant organisations have carried out different activities for soil rehabilitation and conservation including growing Vetiver grass to prevent erosion, promoting organic farming to build up soil organic matter, remediation of abandoned saline soil and soils with other problems, and revising of laws and regulations related to land use (Office of Natural Resources and Environmental Policy and Planning, 2005).

Agriculture and irrigation

The biggest water consuming activity in the Chi River Basin is agriculture, with the largest consumption occurring in the dry season between February and May. Rice is the major crop grown in this area, its production could be increased by improving rice varieties, improving the soils, and increasing the number of crops per year. Some flat areas of the Chi River Basin have very poor soils whose characteristics for rice growing improve when they become anaerobic after being flooded for long periods. Rice yields on these soils significantly increase by adding inorganic fertilizers.

A range of other rainfed crops, such as maize, cassava, sugarcane, and series of smaller crops like soybean, peanuts, and kenaf, are also grown in this area (Nesbitt et al., 2004). In addition, the proportion of other crops is increasing, often with encouragement from the government that is keen to see diversification of agriculture.

Government policies and plans are always driven by economic considerations, but social and environmental issues cannot be neglected. The development of rice irrigation may not be economically profitable, but if it reduces people migration from poor rural areas to large cities, it may have a substantial social benefit as well as a net economic benefit to the country. On the other hand, an irrigation scheme, which may have a net economic benefit, may cause an environmental impact through reducing river flows.

In relation to irrigation, competing users may become an important issue in the Chi River Basin. Large scale irrigation could possibly reduce available water, which may cause insufficient water supplies to provide for both irrigation and domestic needs. Possible solutions could involve the transfer of water from elsewhere in the river basin or by applying a rotation system. Alternatively, there is also some scope to increase wet season irrigation in the Chi River Basin by developing additional small and medium storages, and to otherwise increase water availability by improving irrigation efficiency.

Floods and droughts

Rice growing in paddy fields is the main activity in the Chi River Basin, which is the poorest and least developed area in Thailand. The trends are therefore to develop infrastructure for securing production from floods, to implement irrigation systems to be used during the dry season, and also to provide water storage during the flood periods in order to reduce peak flows downstream.

Floods may be considered as more damaging than beneficial, in particular floods in the Chi River Basin. Urban, semi-urban, and rural areas located along the dike of the Chi River are obviously affected when tremendous floods occur. Whilst the appreciation of flooding in the low plains depends on how land is being used.

Heavy rainfall, which may cause floods near urban areas of Chaiyaphum, Khon Kaen, Maha Sarakham, and Roi Et Provinces, is also a cause for concern in this area. Runoff increases in relation to increase in surface areas. When the water level of the Chi River is high, drainage becomes problematic. This was particularly the case in 2001 with substantial difficulties experienced in Selaphum District of Roi Et Province (Department of Water Resources, 2005b).

With respect to drought, it is a state of water shortage that exacerbates social and economic conditions in the Chi River Basin. These conditions result from the topographical characteristics of the area such that the Southwest monsoon cannot pass through and also from the seasonal variation. Drought situations in the Chi River Basin tend to be progressively more serious. In 2005, the Chi River Basin faced a severe drought due to 2 months delay of the previous year seasonal rainfall, causing water shortage in many reservoirs. The situation was worse due to the increase in water demand from various sectors. Water shortage in some areas was very severe where conflicts regarding water use took place among communities, agricultural, and industrial sectors. Therefore, government has launched urgent and long-term remedies against drought problems. It assigned the Ministry of Natural Resources and Environment, Ministry of Agriculture and Cooperatives, Ministry of Interior, and Ministry of Defence to be in charge of prevention and remedy for urgent drought problems through artificial rain, provision of groundwater, and distribution of water for consumption and cultivation. For long-term remedy, the government established the national strategic water management plan (2004 - 2008) and irrigation systems for sustainable consumption in all sectors. However, an integrated river basin management including public participation could be another option to help solving drought problems (Office of Natural Resources and Environmental Policy and Planning, 2005).

Hydropower development

Like any other aspect of development, the energy demand will also be increased, particularly in the industrial activities. Therefore, the potential for future hydropower development throughout the Chi River Basin has been carried out continuously. However, the chance for such development is highly unlikely to occur due to the limitation of appropriate sites since it will affect a wide range of people and environment, which can lead to stakeholder conflicts. In addition, when considering environmental issues in the past experience with the existing Chulabhorn and Ubol Ratana hydropower dams and reservoirs, perceived that they have caused significant downstream impacts on water quality, and consequently on aquatic biodiversity and fisheries by releasing water with low dissolved oxygen content or even anaerobic water associated with Hydrogen Sulphide (Amornsakchai et al., 2000).

In accordance with the above mentioned constraints, Electricity Generating Authority of Thailand (EGAT), which is the key agency responsible for hydropower development in the river basin, therefore plans to improve the efficiency of the existing schemes in order to serve the increasing demand for electricity by installing electrical generators at Lam Pao Dam, expanding transmission lines, together with improving power interconnections and trade with neighbouring countries (Thai National Mekong Committee, 2004).

Deforestation

Forest cover in the Chi River Basin has significantly decreased (Lorsirirat, 2007). This is consistent with the statement of Boochabun et al. (2007), who stated that between 1952 and 2003 the forest cover has dropped by 20%. Logging, agricultural encroachment, and development of water resources have led to deforestation and forest degradation. The Office of Natural Resources and Environmental Policy and Planning (2005) reported that Royal Forest Department and Department of National Parks, Wildlife and Plant Conservation have launched many significant projects that contribute to achieving objectives of conservation and efficient management of forest resources, in accordance with the sustainable development guideline. To accomplish the objectives, the Royal Forest Department has implemented a four years action plan (2005 - 2008) to manage forest resources by including public participation towards the goal of sustainability. Several projects, namely, One Tambon One Public Park, Promotion of Commercial Forest Plantation, and Forest Plantation Projects have been launched by the department. In addition, the Department of National Parks, Wildlife and Plant Conservation has also implemented a number of projects on Integrated Management of Forest and Wildlife Biodiversity, Management of Protected Areas with Public Participation, Clean Public Parks, and Cheerful Nature.

Wastewater

Due to the growing population in the Chi River Basin, many water supply projects have been developed. Typically, the construction of wastewater treatment facilities did not always catch up with rapid and widespread water supply development. The discharge of untreated or poorly treated wastewater has then caused deterioration of water quality in streams and rivers, particularly downstream of communities, industrial factories and agricultural lands. However, as reported by the Pollution Control Department (2004), the water quality of most of the rivers in the Chi River Basin is still in good condition, except the Pong River, which is in fair condition.

Future of the Chi River Basin

Potential upcoming developments in the Chi River Basin are likely increased agricultural intensification, increased development of hydropower in order to provide adequate electricity within the river basin, and possibly increased industrial development. At present, the main focus of agricultural intensification is to increase irrigation in order to allow additional crops during the dry season. In the short-term, this will depend either on pumping water from groundwater resources or directly pumping water from the river. In the longer term, it could also involve construction of additional storage reservoirs, weirs, and dams.

Agriculture in Chi River Basin is not intensively developed throughout much of the river basin, it remains more on extensive development. However, the portion of the river basin with the most intensive agricultural development is also the portion, which is most suffering from the impacts of severe environmental problems. Those problems, which are not caused by developments or activities upstream, are locally generated.
Therefore, it is time to find a better way to boost up agricultural productivity in the way towards more efficient water management and use. Moreover, it is quite clear that discussions on 'how hydro and non-hydro electricity sources are developed and managed' is of particular interest in case there is an attempt to address the issue more wisely than what has been done in the past.

3 Water resources and land use pattern in the Chi River Basin

3.1 Land and water resources

Land and water resources are the most valuable natural resources that are essential for the existence of life. However, there is evidence that conflicts over land and water resources in the Chi River Basin are happening more and more frequently. This situation is aggravated by the fact that there are many different users and overlapping uses. In other words, the needs of agriculture, commerce, industry, domestic and others often result in diversion from one use to the other. To tackle the problems, the totality of land and water interactions has to be synergised through appropriate policies.

3.1.1 Land resources

According to Eswaran et al. (2000), land is a complex system that encompasses both biophysical and socio-economic aspects and the interaction of these characteristics results in a multitude of entities. There is a lot of pressure on land-related resources due to the increase in population accompanied by greater economic and social exploitative demands. To determine the land suitability/limitations and comply with other resources for optimal level of management, a number of aspects are considered and described in the following details.

- *land morphological characteristics.* On the basis of morphological characteristics, land morphology in the Chi River Basin can be distinguished into five main types:
  - floodplain, the land along the river banks where it commonly floods during the rainy season. The soil has high clay content and poor drainage, it is often found in low-lying areas and mainly planted with rice. In high areas along the rivers, the soil is mainly of loam texture with moderate drainage and largely utilized for field crops and settlements, while rice is still grown in some areas;
  - low and semi-recent river terrace, low-lying areas along the banks of the river where rice is grown primarily. The most common soil in the area is sandy loam and exhibits poor drainage characteristics, under waterlogged conditions at 20 to 30 cm depth;
  - middle to high terrace, the land surface is relatively smooth and mainly characterized by undulating topography. The soil textures include sandy loam and clay loam, and are largely exploited for planting field crops and fruit crops;
  - erosion surface and footslope, it is characterized by an undulating to rolling topography. The soil is partly sandy loam, which displays good drainage conditions, partly falls into the skeletal soil category, and partly the parent material layer at the level from 60 to 100 cm below the ground surface. The land is mainly planted with field crops and fruit crops, while some areas still remain under forest;
  - hilly and mountainous area, the areas with significant elevation differences that fall under slopes greater than 30 to 50%, and most of the area is still under forest cover.

- *land suitability classification.* In essence, a detailed assessment of the land resources based on the land suitability can provide opportunities to manage land for the best purposes or make land suitable for a defined use or practice. The analysis was undertaken to characterize and evaluate the land suitability for two specific uses, i.e. suitability for cash crops and irrigation, which can be summarized as follows:
  - suitability for cash crops, i.e. rice, field crops, and fruit crops. From a wide variety of landscape morphological characteristics, the analysis is primarily based upon physical and chemical soil characteristics to classify land according to its suitability for cash crops. The results are presented in five basic suitability classes: suitable, moderately suitable, marginally suitable, not suitable, and miscellaneous land. Table 3.1 shows a summary of the land suitability classification presented in percentage of total area for each category;

Table 3.1 Land suitability for cash crops

Suitability	Area in 1,000 ha	Area in %
For rice:		
- suitable	566	11.5
- moderately suitable	398	8.1
- marginally suitable	708	14.3
- not suitable	2,651	53.7
- miscellaneous land	614	12.4
For field crops:		
- suitable	300	6.1
- moderately suitable	1,268	25.7
- marginally suitable	1,002	20.3
- not suitable	1,753	35.5
- miscellaneous land	614	12.4
For fruit crops:		
- suitable	897	18.2
- moderately suitable	809	16.4
- marginally suitable	865	17.5
- not suitable	1,753	35.5
- miscellaneous land	614	12.4

Source: Department of Water Resources (2006)

  - suitability for irrigation. In deciding upon the land suitability for irrigation in addition to the degree of soil suitability, consideration is given to both economic (production cost and yield potentials, land development cost, etc.) and physical conditions (topography, slope, relief, drainage, hydrology, climatic characteristics, etc.) (Bureau of Reclamation, 1951). In the potentially irrigated land, the suitability for irrigation was evaluated by applying the Land Classification System, developed by the Bureau of Reclamation of the U.S. Department of the Interior. Six land classes are recognized in the Chi River Basin, as shown in Table 3.2.

Table 3.2 Percent distribution of land potentially suitable for irrigation

Suitability	Potentially suitable land	Area	
		in 1,000 ha	in %
1R	Lands that are highly suitable for rice cultivation under irrigation	750	15.1
2R	Lands that have moderate to fair suitability for rice cultivation under irrigation	1,003	20.3
C1	Diversified croplands that are highly suited to irrigation farming	82	1.7
C2	Diversified croplands that have moderate to fair suitability for irrigation farming	1,472	29.8
C5	Lands that are marginally suitable for irrigated farming	322	6.5
C6	Non-arable, land not suited to irrigated farming	723	14.6
SC	Mountain	489	9.9
U	Residence	4	0.1
W	Waterbody	102	2.0
	Total	4,947	100.0

Source: Department of Water Resources (2006)

- *soil salinity and its distribution.* Based on the classification of salt affected soils, few salinity problems in the Chi River Basin are evident, but 99.9% of the area is having less than 50% salt crust coverage. The areas under slightly salt-affected soils, i.e. salt crust percentage on soil surface less than 1%, are mostly found in the southern part, while soils with high salinity levels (areas covered between 10 and 50% by salt crust) are prevalent in the southwestern part. Of the salt affected soils listed in Table 3.3, lands can be placed into eight different salinity categories.

Table 3.3 Soil salinity levels and extent of salt-affected soils in the Chi River Basin

Salinity category	Salt crust on soil surface in %	Area	
		in 1,000 ha	in %
Very severely salt affected areas	> 50	4	0.1
Severely salt affected areas	10 - 50	13	0.3
Moderately salt affected areas	1 - 10	158	3.2
Slightly salt affected areas	< 1	998	20.2
Non salt-affected areas	-	2,479	50.0
Rock salt underlying below the land surface	-	686	13.9
Mountain	-	521	10.5
Waterbody	-	89	1.8
Total		4,947	100.0

Source: Department of Water Resources (2006)

- *soil erosion assessment.* As a result of the assessment of soil erosion under the current condition for the Chi River Basin by employing the Universal Soil Loss Equation (USLE) within a Geographic Information System (GIS) environment, the findings confirm that the overall erosion rate is low, although parts of the river basin are steep. However, since the proportion of land area is covered by forest, soil erosion is being reduced. Lands previously used for shifting cultivation also result in high soil erosion, if they are in the upper hilly regions, as well as in upland agricultural areas. The results of the analysis for annual soil loss rates for the Chi River Basin are shown in Table 3.4.

Table 3.4 Annual soil loss rates for the Chi River Basin

Evaluation class	Erosion rate in ton/ha/year	Area in %
Very mild	0.0 - 12.5	85.3
Mild	12.5 - 31.3	3.8
Moderate	31.3 - 93.8	4.1
Severe	93.8 - 125.0	0.8
Very severe	> 125.0	6.0
	Total	100.0

Source: Department of Water Resources (2006)

3.1.2 Water resources

Surface water resources

The average annual flow in the Chi River system is 11,900 million m^3/year or 242 mm/year runoff depth over the entire Chi River Basin, while the annual rate of runoff per unit area is 7.64 l/s/km^2. The river carries 10,300 million m^3/year or 86% of its average annual flow during the rainy season, and 1,600 million m^3/year or 14% in the dry season. The maximum discharge of the river observed in September was 3,300 million m^3/year or equivalent to 30% of the total annual flow as shown in Figure 3.1.

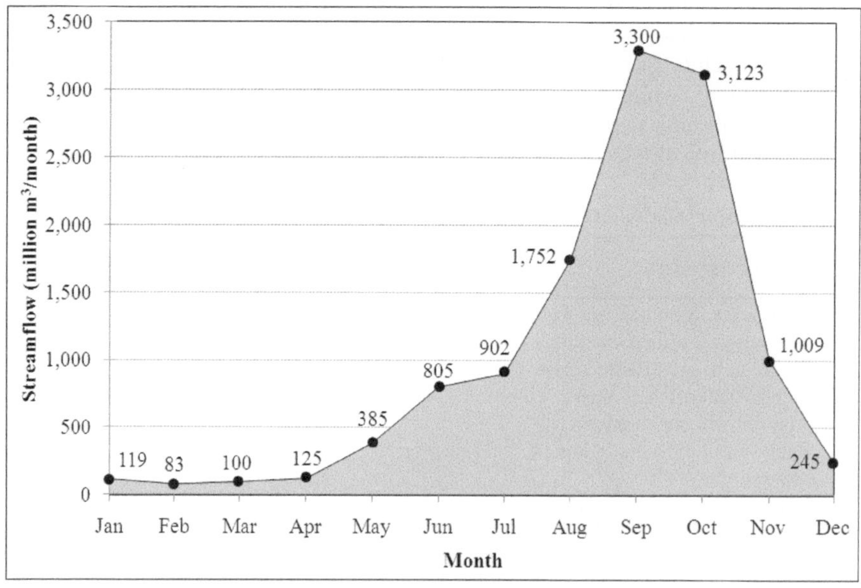

Figure 3.1 Average monthly streamflow of the Chi River Basin for the period 1973 - 2002 (Department of Water Resources, 2006)

A large number of water resource development schemes took place in the Chi River Basin in response to the rapid increase in water demand. In order to reach the consumptive demand, the Royal Irrigation Department maintains and operates approximately 1,868 schemes, which include large, medium and small scale reservoirs, and electrical pumping schemes (Table 3.5). A total storage capacity of 4,970 million m^3 has been created in the river basin through development schemes, and some 536,000 ha of land are being irrigated.

Table 3.5 Summary of water resource development schemes (adapted from Royal Irrigation Department (2008))

Scheme	No. of schemes	Storage capacity in MCM	Irrigated area in 1,000 ha
Large scale	6	4,017	132
Medium scale	96	766	133
Small scale	1,319	187	100
Electrical pumping	447	-	171
Total	1,868	4,970	536

Groundwater resources

Groundwater quantity and quality vary considerably across the Chi River Basin, largely related to the geology of the aquifer system. The groundwater flow will take place preferably at the opened-up joints and fractures of consolidated rock. In the permeable parts of the fracture zone, the groundwater flow passages are located, while the other parts with less fracture, the groundwater flow can play only an insignificant role or even no groundwater flow takes place. Moreover, according to the rock salt structure, which lies on the underlying Maha Sarakham Formation, this will provide opportunity for saline water to be drawn upwards. Based on the findings from assessment of groundwater potential, the Chi River Basin has an estimated groundwater storage volume of 5,010 million m^3, about 7.13% of the total river basin area has yield greater than 20 m^3/hr (Figure 3.2). However, it can only withdraw a total of 85 million m^3/year, or approximately 1.7% of the total groundwater resources. At present, a groundwater volume of 70 million m^3 could be extracted annually to serve domestic and industrial needs.

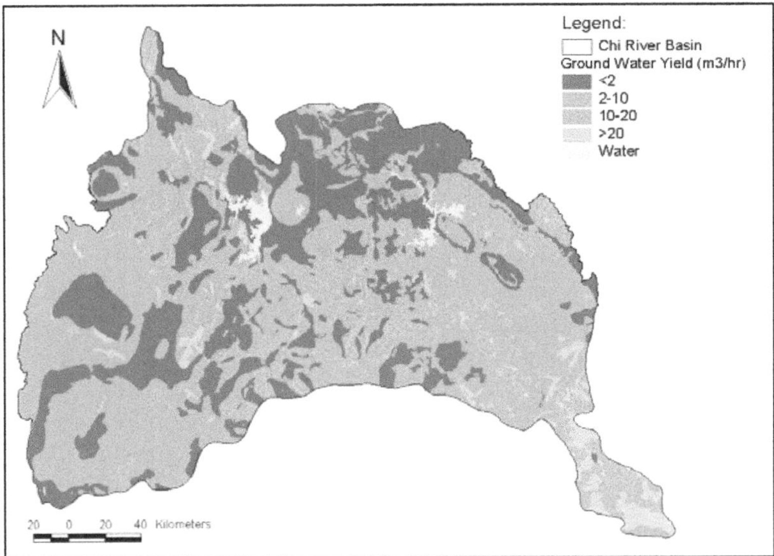

Figure 3.2 Groundwater yields in the Chi River Basin (Regional Centre for Geo - Informatics and Space Technology, 2001)

Water demand

The 2002 water demand in the Chi River Basin is estimated to have been 4,210 million m^3/year, however, the demand is expected to increase significantly. Water demand can be broken down into three major categories as briefly described below:

- *domestic demand.* The actual demand served was approximately 188 million m^3/year, representing for some 4.4% of the total water demand. Of this total, 32 million m^3/year is served by groundwater;
- *industrial demand.* The water demand for this sector was estimated at 308 million m^3/year. Whilst this activity utilized 38 million m^3/year of groundwater;
- *agricultural demand.* Demand in agricultural water utilization from a production perspective was roughly estimated at 3,720 million m^3/year. It is obvious that this sector plays a crucial role in total demand. Within this demand, water use covers the following components:
    - water demand in irrigated areas, water consumption in this sector is about 3,090 million m^3/year, which constitutes 73.4% of the total water demand;
    - water demand in non-irrigated areas, concerns the need for water along the rivers. The demand side receives and distributes the water by means of electrical pumping units from the rivers, excluding rainfed agricultural area. The amount of water used was in the order of 625 million m^3/year, which concerned 14.8% of the total water use.

Moreover, a water demand for generating electricity and the downstream river ecosystem does exist in this river basin, which amounts to about 3,230 million m^3/year. However, its flow is relatively unaffected by water abstraction.

3.2 Hydrology

3.2.1 Climate and rainfall

The climate of the Chi River Basin is basically seasonal in nature. Moreover, the climatic conditions often result in floods and droughts. To depict the distribution and variation of climatic elements, all relevant climate information has been taken from the Thai Meteorological Department (TMD), which were obtained from 7 meteorological stations located inside the Chi River Basin and its vicinity. The stations are: Khon Kaen, Chaiyaphum, Kosum Phisai, Roi Et, Ubon Ratchathani, Udon Thani, and Wichian Buri.

The Chi River Basin is located in the tropical monsoon region, i.e. the paths of the two branches of the southwest and northeast monsoons. The southwest or summer monsoon brings prevailing winds that cause heavy rainfall over the entire area. In May, the rainfall rises steadily until a maximum is reached in August or September, and declines rapidly in late October or early November. The northeast or winter monsoon starts in November and ends in February. The weather is generally dry with occasional light showers. In addition to being relatively cool and dry months, December and January tend to be the coolest and driest months, and the hottest month is April.

To get a good insight into the climatic conditions of the Chi River Basin, each of the climate parameter has been broken down into several individual variables as shown in Table 3.6.

The Chi River system is largely dependent on the monsoon rains. Average annual rainfall is 1,170 mm/year. About 89% of the annual rainfall comes during the rainy season (May to October), while the rest 11% accounts for the dry period (November to April). Moreover, it is also found that the amount of rainfall that falls over the Chi River Basin is irregular through the months as can be seen in Figure 3.3.

Table 3.6 Summary climate information

Climate variables*	Mean annual values	Unit
Temperature	27.0	°C
Humidity	71.3	%
Wind	2.2	knots
Cloudiness	5.5	Octa
Class-A pan evaporation**	1,771	mm/year
Potential evapotranspiration	1,824	mm/year

Note: * Data are mean annual values from a 30-year record (1971 - 2000)
** An estimate of open water evaporation is obtained by multiplying a pan coefficient of 0.7
Source: Royal Irrigation Department (2008)

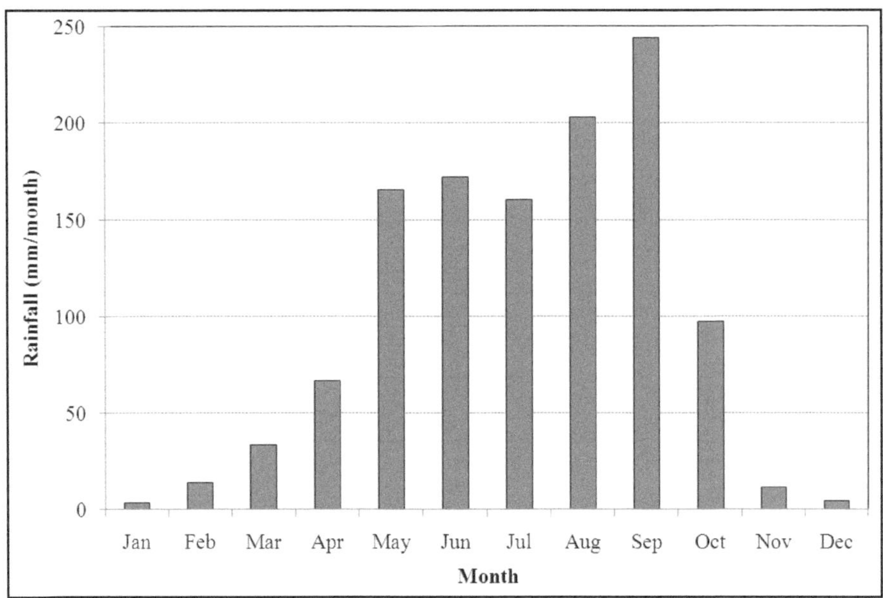

Figure 3.3 Distribution of average monthly rainfall in the Chi River Basin (Royal Irrigation Department, 2006)

3.2.2 River system

The Chi River originates from the Phetchabun mountain range, flows eastward and joins the Mun River at about 100 km upstream of the confluence with the Mekong River. It is the main waterway flowing through the central part of the northeast region of Thailand. According to the differences in relief terrain, the rivers are steep in the upper reaches and fairly steep in the middle reaches, while the lower reaches flow through the relatively flat gradients.

From the flow contribution point of view, the Chi River system consists of 10 principal tributaries, relevant particulars of which are given in Table 3.7.

Table 3.7 Details of the principal tributaries of the Chi River

Name of the tributary	Sub-basin area in ha	Length in km
Lam Nam Phong	638,800	273
Lam Nam Phrom	225,400	195
Lam Nam Choen	293,200	129
Lam Phaniang	187,500	185

Table 3.7 Details of the principal tributaries of the Chi River (cont'd)

Name of the tributary	Sub-basin area in ha	Length in km
Huai Sai Bat	67,600	75
Lam Pao	746,700	199
Lam Phan Chart	69,600	71
Lam Khan Chu	173,100	97
Lam Krachuan	89,300	78
Lam Nam Yang	413,400	217

Source: Royal Irrigation Department (2008)

To solve water resources allocation problems, large-scale water resources systems are enabled capturing water from the upper part of the Chi River Basin. Consequently, reservoirs were created by dams that were designed and built and to serve several purposes, as shown the detail in Table 3.8.

Table 3.8 Major reservoirs in the Chi River Basin

Name	Sub-basin/Province	Primary purpose	Storage capacity in MCM	Catchment area in ha
Ubol Ratana	Upper Phong/ Khon Kaen	- Irrigation - Domestic - Hydropower	2,263	1,200,000
Lam Pao	Lower Lam Pao/ Kalasin	- Irrigation - Domestic - Flood control	1,430	596,000
Chulabhorn	Upper Phrom/ Chaiyaphum	- Irrigation - Hydropower	188	54,500

Sources: Royal Irrigation Department (2008)
Electricity Generating Authority of Thailand

3.3 Water management and flood protection schemes

3.3.1 Opening

In the face of water vulnerability, one might consider the water management related issues with the intention of finding ways to manage water resources more wisely, efficiently, and beneficially. However, not only problems of water management that need to be addressed in a water perspective, but other interests such as flood protection, which is in fact relevant to water resources management should also be included. In other words, emphasis has to be placed on both aspects: water scarcity and floods, which are of major challenges for water resources development and management in the Chi River Basin. However, a number of water management and flood protection schemes still have not been able to achieve the desired objectives as a result of several factors. One of those factors is the inappropriate institutional framework, which should ideally be taken into consideration. Notwithstanding, the integration of structural and institutional measures for flood protection and water management is widely believed to be a key requirement for tackling these water-related problems.

3.3.2 Water management and land use under different water use practices

Water management and land utilization under different water use practices play a significant role in natural resources management. However, the quantity and quality of land and water resources in the Chi River Basin have been deteriorating due to encroachment and destruction of the river basin area. Currently, it faces two major problems, i.e. flooding and severe water shortages in some regions and in coming decades these constraints are expected to become more serious and widespread due to an increase in water demand. Thus, there is a need to search for a balance between water resources management and land utilization in order to achieve more sustainable ways of working with natural systems.

Water management

Over the past years, there were abundant water resources in the Chi River Basin, however, convenient access to water often led to waste of water resources. In brief, not much attention was paid to efficient water resources management, as well as an effective plan was absent. It appears further that increasing population density and intensity of economic activity are impeding efficient management of water resources. Hence, there is an urgent need for judicious water allocation management.

As societies develop, water resources are increasingly utilized for several purposes. Yet the Chi River Basin still struggles to allocate sufficient water resources to meet those needs. As a result, competition for water exists between regions, between different sectors, and between upstream and downstream users in the river basin. There is also a competition of ideas over the most appropriate ways to manage, govern and allocate water in the Chi River Basin. To overcome these shortcomings and to meet the rising demand for water, management of water resources needs to form a wide range of development endeavours. Table 3.9 shows the main features and water use per economic sector in the Chi River Basin based on the data of 2002.

Table 3.9 Main features of the Chi River Basin

Main features	Unit	Total
Drained area	ha	4,950,000
Population density	person/km^2	134
Storage capacity	Mm^3	5,000
Irrigated area	ha	536,000
Water use	Mm^3/year	
- Irrigation		3,720
- Domestic		188
- Industry		308
- Hydropower		1,860
- Ecological balance*		1,370
Total		7,440

Note: * A minimum flow required to maintain a natural and ecological balance during the usage of water in all its dimensions

Source: Department of Water Resources (2006)

Land utilization under different water use practices

With respect to the land utilization, each physiographic area can be different according to their dominant land use patterns. In terms of land utilization in relation to physiography in the Chi River Basin, there are also several types, which can be described as follows.

The hilly and mountainous areas of the Chi River Basin are mostly accounted for trees and field crops. Undulating or rolling landforms accommodate both long duration field crops (mainly cassava and sugarcane) in the upland cultivated areas, and wet season paddy in the lowland. However, low soil fertility and erratic rainfall with a long drought period, as well as flooding, are still the main problems as it could limit farming potential.

The non-floodplains, which are primarily covered by sandy loam and to a lesser degree by loamy clay, are generally under rice cultivation in the rainy season. If capillary soil water is available, a second crop of peanut, soybean, water melon, etc., will be cultivated after the wet season.

As the economy of the Chi River Basin is predominantly agricultural, it plays a key role in economic development of the river basin. Therefore, attention is devoted primarily to the agricultural land utilization.

In Chi River Basin, the major cultivation systems have been categorized based on climatic conditions and management practices. Two systems, i.e. rainfed and irrigated cultivation are widely known forms of cultivation practices. Moreover, human exploitation under different forms of agricultural practices can also be remarkably divided into two main classes: upland and lowland. More details about land and water utilization will be given by the following:

- *rainfed cultivation.* Cropping systems in rainfed cropland are divided into 12 broad categories and most of them are based on one crop per year (Table 3.10).

Table 3.10 Cropping system in rainfed cropland areas

Area	Cropping system
Lowland	- Rice (during the rainy season and only one crop) - Rice-Vegetables - Rice-Off-season rice - Rice-Beans - Rice-Waxy corn/Sweet corn - Rice-Watermelon
Upland	- Sugarcane - Fodder corn-Cassava - Fodder corn - Cassava - Perennial plants - Fruit crops

Source: Department of Water Resources (2006)

- *irrigated cultivation.* Cropping systems on irrigated land are commonly divided into thirteen categories under different crop rotations and management regimes (Table 3.11).

Table 3.11 Cropping system in irrigated cropland areas

Area	Cropping system
Lowland	- Rice - Rice-Off-season rice - Rice-Beans - Rice-Waxy corn/Sweet corn - Rice-Vegetables - Rice-Tomato - Rice-Watermelon
Upland	- Fodder corn - Sugarcane - Soybean

Table 3.11 Cropping system in irrigated cropland areas (cont'd)

Area	Cropping system
Upland	- Field crop (i.e. kenaf) - Perennial plants - Fruit crops (i.e. mango)

Source: Department of Water Resources (2006)

- *cropping calendar*. By definition, the cropping calendar is therefore designed to specify the times of different stages of the growing season. Throughout the Chi River Basin, farmers usually adopt a similar cropping calendar as shown in Figure 3.4. For instance, lowland rice cultivation usually starts between July and August, and the harvest is usually gathered towards the end of the year, mainly in December. During the dry season, irrigated cultivation is proposed to start between December and January and end by April. In uplands, the first season field crop (i.e. fodder corn) is planted in May and harvested in July, while the second season field crop is sown in August and harvested during October and November. In light of this, an adjustment of the cropping sequence is necessary to more closely match the anticipated season.

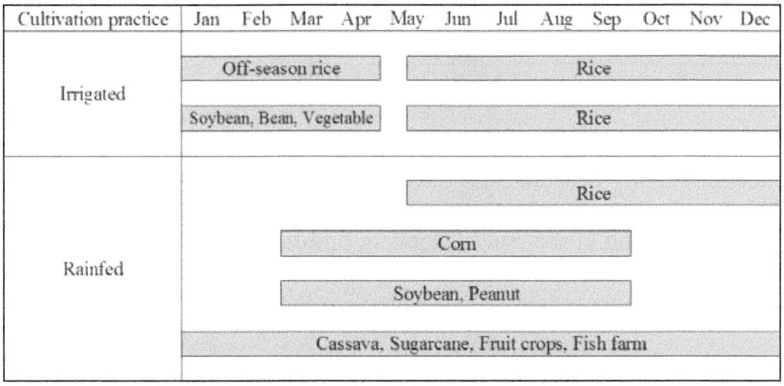

Cultivation practice | Jan Feb Mar Apr May Jun Jul Aug Sep Oct Nov Dec

Irrigated: Off-season rice; Rice; Soybean, Bean, Vegetable; Rice

Rainfed: Rice; Corn; Soybean, Peanut; Cassava, Sugarcane, Fruit crops, Fish farm

Figure 3.4 Cropping calendar used in the Chi River Basin (Department of Water Resources, 2006)

3.4 Flood management and irrigation systems

3.4.1 Preface

For sustainable development of the Chi River Basin, it is a prerequisite to achieve a comprehensive flood management plan to increase preparedness for recurrent floods.

The flood events in the Chi River Basin have revealed considerable weaknesses in the way flood calamities are being addressed. Information on the extent of the flood-affected areas and the extent of the damages are still not well known and collected data show conflicting information. The unpreparedness and lack of procedures to assess the extent and damage caused by flood damage has affected the allocation and mobilization of emergency assistance and the readiness of local institutes and agencies to offer effective support. Therefore, a better understanding of the dynamics of floods is needed by developing procedures to better assess flood behaviour and defining various answers to prevent and restrict the damage caused by floods. As a result, optimal solutions can be recommended to overcome the effects of the floods and to address in a more sustainable way the recurrent effects of floods.

With respect to the irrigation systems, irrigation is an integral part of the agricultural production systems. As population in the region is growing rapidly, irrigation is important for food security as well as for economic development.

3.4.2 Flood prone zone

To take necessary preparations prior to the arrival of the flood, it is essential to carefully identify the flood prone zones within the borders of the Chi River Basin as the flood prone residents always overlook and develop confidence in the fact that future flooding would not extend beyond what currently exists.

In order to address a flood prone area of the Chi River Basin, the delineation of flood prone zone boundaries was then created with various hazard categories, as given in Table 3.12.

Table 3.12 Areal estimates of different flood prone areas in the Chi River Basin

Flood prone zone	Area	
	in 1,000 ha	in %
High risk	376	7.6
Moderate risk	1,013	20.4
Low risk	137	2.8
No risk	3,315	67.0
Waterbody	107	2.2
Total	4,948	100.0

Source: Department of Water Resources (2006)

It has been found that the high risk flood prone area is mainly located along the floodplains of the Chi River and its major tributaries (Figure 3.5). There are a number of factors contributing to the crucial flooding problem in these flood prone zones such as low topographic relief landscapes, flat slope and poor features of the drainage systems, high intensity rainfall, and high runoff volume and peak rates.

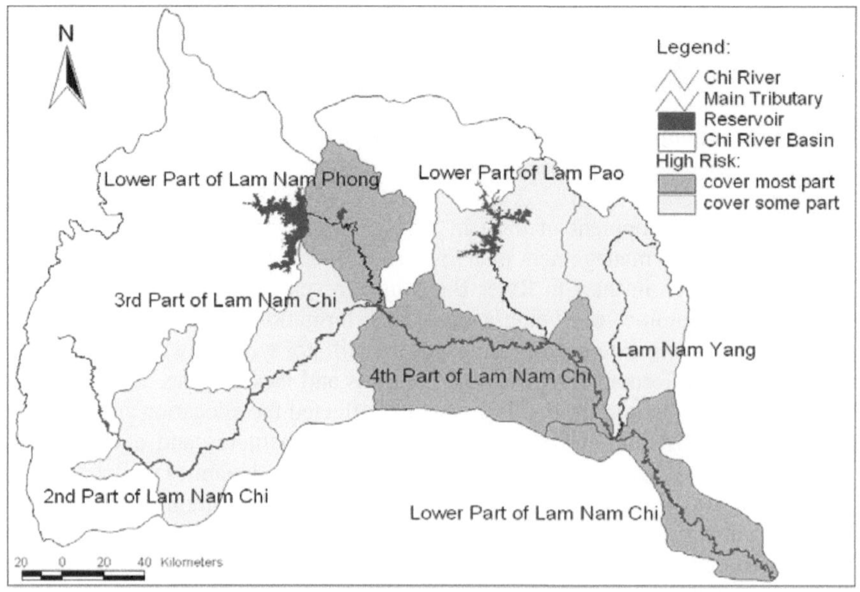

Figure 3.5 High risk flood prone area in the Chi River Basin

The most flood prone area covers most parts of the following sub-basins, i.e. Lower Part of Lam Nam Phong, Fourth Part of Lam Nam Chi, and Lower part of Lam Nam Chi. Moreover, the high risk flood prone area is also partially located in some sub-basins such as Lower Part of Lam Pao, Lam Nam Yang, and second and third parts of Lam Nam Chi.

3.4.3 Types of flood management

According to the experiences from catastrophic flood events, the capability of the existing flood control system is limited because of the large volumes of floodwater generated by the Chi River Basin. To reduce the destructive effects of floods, there are a variety of structural flood management measures implemented over the past decades to keep floods away from people in the Chi River Basin. In general, those existing flood management measures have sometimes been successful in such a way that the flood characteristics (e.g. discharge, level) of the rivers do not exceed the design parameters of a particular flood mitigation structure. Indeed, they are able to regulate the impacts and reduce the damages and losses due to floods to a certain extent, but it is absolutely impossible to achieve an entirely complete protection. With respect to the implementation of flood management works, types of flood management in the Chi River Basin can be distinguished as follows.

Storage in reservoirs

The existing reservoirs, i.e. Ubol Ratana and Lam Pao, have the potential controls at approximately 19% and 12% of the flow regime in the Chi River Basin, respectively, which was calculated based on data from the Royal Irrigation Department (2008). However, due to the excessively large spillway design flood, the flood control capabilities of both reservoirs are limited. Hence, capacities of spillway and flood storage above the crest of the spillways of both reservoirs are also limited. As a result, there is no opportunity for these two reservoirs to provide flood storage for regular floods.

Ubol Ratana reservoir is the largest storage in the Chi River system. With its full capacity, it represents about one fifth of the total flow regime of the river basin. Due to this storage effect, attenuation of the inflow hydrograph can be obtained by operating the reservoir according to a design rule curve. In such situation, effective control can provide a significant flood storage volume, but at the same time it will limit the potential flood control benefits.

A more thorough consideration of Lam Pao reservoir, the operation results in a significant reduction in outflow, which will have a beneficial effect on areas downstream. To maximize the flood damage reduction benefits to downstream areas, any attempt to adjust the reservoir rule curve would be required whereby the reservoir level is deliberately lowered during the wet season to accommodate anticipated flood inflows. In doing so, a significant flood control could only be achieved at the expense of irrigation security, i.e. there may not be sufficient water in the storage to meet water demand for irrigation during the dry season.

Briefly, the major constraint to both Ubol Ratana and Lam Pao reservoirs is related to the limitation of the spillway capacity in combination with the restriction of river offtake facilities compared to reservoir capacity. When flooding of great magnitude occurs, it can result in a large rise upstream level at each reservoir. Based on these findings, a number of remedial actions, which resulted in modifications of both reservoirs, were proposed and subsequently implemented. That is to say, for Ubol Ratana reservoir, the crest of the dam was raised by 3.1 m, while the spillway was lowered by 5 m with the change of its operation system from open chute to orifice controlled by radial gates (Electricity Generating Authority of Thailand, 1997). In the case of Lam Pao reservoir, the new primary spillway controlled by radial gates was constructed, while the previous ogee type spillway intake is adjusted and used as an emergency spillway (Royal Irrigation Department, 2009).

Flood protection dike

Beyond primary storage reservoirs, flood mitigation can be a measure in the Chi River Basin. It is also referred to building of dikes as they are generally used to protect urban areas and agricultural land from relatively frequent floods. Conforming to the report of the Royal Irrigation Department (1988), a significant number of flood dikes have been constructed in some parts of the Chi River and its main tributaries, i.e. Nam Phong River and Lam Pao River, with a total length of approximately 300 km. These defences, built by the Royal Irrigation Department in the 1950's, 1970's, and 1980's, were mostly designed to provide protection against a flood with a 10% annual probability of exceedance. The different aspects of the schemes are discussed in more detail below and are illustrated in Figure 3.6.

Nong Wai and Nam Phong schemes

The Nong Wai scheme is located on the right bank of the Nam Phong River and on the left bank of the Chi River. These dikes were originally constructed in 1979 and have been improved to the 10% annual probability of exceedance protection level since 1983.

The Nam Phong scheme is situated on the left bank of the Nam Phong River down to its confluence with the Chi River, some 70 km along the left bank of the Chi River down to the confluence with the Huai Chiang Song River as well as on the right bank of the Huai Chiang Song River. These dikes were completed in 1985 with the protection level of 10% annual probability of exceedance.

Lam Pao scheme

The Lam Pao scheme area is divided into two separate stages, i.e. the Lam Pao Stage I and the Lam Pao Stage II.

The Lam Pao Stage I consisted of building flood protection dikes on the left bank of the Lam Phan River, and on the right bank in the middle reach of the Lam Pao River.

The Lam Pao Stage II involved the construction of a 53.3 km-long dike on the left bank of the Chi River, and a 22.4 km-long dike on the right bank of the lower reach of the Lam Pao River. The construction of both dikes was completed in 1985 and provides a level of protection against 10% annual probability of exceedance flood.

Ban Tum - Ban Tew scheme

The Ban Tum - Ban Tew scheme stands on the right bank of the Chi River between the Maha Sarakham Municipality area and the Chi-Huai Khakhang confluence. The 26.5 km-long dike with 10 outflow gates, which can protect up to 4,600 ha of agricultural land and communities from flooding, was completed in 1955.

The scheme was recently extended 12 km upstream on purpose to protect the Maha Sarakham Municipality area from flooding, which also resulted in an increased area for agricultural cultivation.

The original portion of the Ban Tum - Ban Tew flood dike scheme was upgraded in 1986 by using the record of flood level in 1980 at Ban Tha Khon Yang gauged site (E8A) as the criteria (also for the design of the extended dike).

Thung Saeng Badan scheme

The Thung Saeng Badan scheme was constructed over the past 50 years, extending nearly 60 km along the right bank of the Chi River, whereby the improvement works were completed in 1986. The protection level is currently defined with 10% annual probability of exceedance.

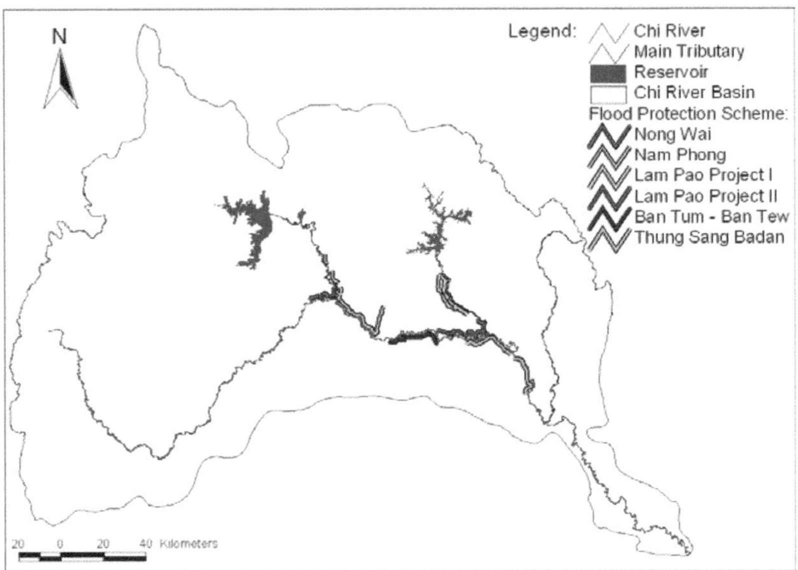

Figure 3.6 Existing flood protection dikes in the Chi River Basin (Royal Irrigation Department, 1999)

3.4.4 Types of irrigation systems

During the last half century, numerous irrigation development schemes have been implemented in the Chi River Basin, as a result of a shortage of water during the dry season. A number of dams and reservoirs have been developed for irrigation, hydropower, industry, flood control, water supply, navigation, and fisheries.

Consequently, irrigation and hydropower schemes are the important users of water in the Chi River Basin. Basically, irrigation is providing supplementary water in the wet season and the total requirement in the dry season. At present, approximately 536,000 ha of agricultural land is irrigated (Royal Irrigation Department, 2008). In order to classify the different development schemes, the Royal Irrigation Department has been categorising the irrigation development by setting the classification criteria for different scale of irrigation systems by considering the technical parameters, i.e. storage capacity, water surface area, and irrigated area (Table 3.13).

Table 3.13 Classification criteria for large-, medium-, and small-scale irrigation schemes

Criteria	Types of irrigation schemes		
	Large-Scale	Medium-Scale	Small-Scale
Storage capacity	>100 Mm^3	2 - 100 Mm^3	<2 Mm^3
Water surface area	>15km^2	<15km^2	-
Irrigated area	>12,800 ha	480 - 12,800 ha	<480 ha

Source: Royal Irrigation Department (2008)

The large-scale gravity schemes are often situated in the valleys of the main rivers. Water supply is from a network of canals abstracting water either directly from upstream reservoirs or from diversion structures situated on the river channel below the reservoir. Return flows and drainage may be put back into the main river channel many kilometres downstream of the abstraction point. Some characteristics of the large-scale water resources development schemes are presented in Table 3.14.

Table 3.14 Large-scale water control structures in the Chi River Basin

Name	Sub-basin/Province	Storage capacity in MCM	Irrigated area in 1,000 ha
Ubol Ratana Dam[1]	Upper Phong/Khon Kaen	2,263	-
Lam Pao Dam[2]	Lower Lam Pao/Kalasin	1,430	50
Chulabhorn Dam[1]	Upper Phrom/Chaiyaphum	188	-
Kumphawaphi Weir[2]	Upper Lam Pao/Udon Thani	102	9
Nong Wai Weir[2]	Lower Phong/Khon Kaen	34	42
Thung Saeng Badan Weir[2]	IV part of Chi/Roi Et	-	31
Total		4,017	132

Note: [1] Under the responsibility of the Electricity Generating Authority of Thailand
[2] Under the responsibility of the Royal Irrigation Department, Thailand

Sources: Department of Water Resources (2006)
Royal Irrigation Department (2008)

Typical medium-scale schemes consist of small reservoirs or water tanks. Tributary runoff is impounded in small reservoirs and released for irrigation by gravity to areas immediately downstream of the storage.

Small-scale schemes, which are mainly for domestic water consumption, small informal irrigation, and livestock, ranging from simple diversion weirs directly to village water tanks.

In addition, water can also be delivered by centrifugal or axial pumps, where water is abstracted or diverted directly from a river channel up to the steep river banks. The pumped water then flows under gravity through a network of small canals into the scheme.

3.5 Institutional arrangements with respect to water management and flood protection

The Chi River Basin is faced with water-related problems such as water shortage, floods and poor water quality. These problems have caused severe damage and adverse impacts on the economy and people's standard of living.

Previously, tailoring efforts focused on constructing new water development projects rather than on water management (Suiadee, 2002). The situation even got worse since the authorities did not provide water users with any opportunity to take part in the decision-making procedures, in which they were supposed to be involved in the plan development, maintenance, or even day to day water management. As a result, any operational activity with respect to water management, which affects a lot of stakeholders with different interests, can lead to a conflict regarding the distribution of water. Therefore, to get rid of the constraints, a water management approach has to be developed towards active involvement of the water-sharing stakeholders at every stage of the development process. After that a river basin-wide integrated plan would be needed to comprehensively cover the aspects of managing water resources and other relevant resources in the Chi River Basin in order to deal with the recurring catastrophes. In addition, an institutional framework, which will help to shape exposure, sensitivity and capacities to respond of individuals, social groups and social-ecological systems, also needs to be strengthened (Lebel et al., 2011).

3.5.1 Institutional arrangements for water management in Thailand

Water resources planning, development and management is critical to sustain future economic growth in Thailand. Water resources play and will continue to play a fundamental role in meeting the growing demand for domestic water consumption, agricultural and industrial production, hydropower and tourism sectors.

During the critical periods such as floods and droughts in the river basins, the Cabinet usually sets up an ad hoc committee to manage the water resources for the benefit of all users in the river basins. The National Water Resources Committee (NWRC) is the secretariat office, which has the responsibility of coordinating with other agencies concerned.

3.5.2 Agencies involved

There are 30 executing and implementing agencies from 8 ministries, 5 state enterprises and 2 independent public agencies nationwide working in water resources development and management. In principle, each of the agencies has its own roles and responsibilities, in which some of them are shared responsibilities among different agencies. The list of the mentioned agencies is as follows (House of Representatives' Ad Hoc Committee on Solutions to Water Resources, 2008):

- The Prime Minister's Office: Office of the National Economic and Social Development Board (NESDB), Bureau of the Budget (BB), Office of the Civil Service Commission (OCSC);
- Ministry of Agriculture and Cooperatives: Office of the Permanent Secretary for Agricultural and Cooperatives (OPSMOAC), Land Development Department (LDD), Department of Fisheries (DOF), Agricultural Land Reform Office (ALRO), Royal Irrigation Department (RID), Cooperative Promotion Department (CPD);
- Ministry of Interior: Department of Disaster Prevention and Mitigation (DPM), Department of Provincial Administration (DOPA), Department of Public Works and Town & Country Planning (DPT);
- Ministry of Natural Resources and Environment: Office of Natural Resources and Environmental Policy and Planning (ONEP), Pollution Control Department (PCD), Royal Forest Department (RFD), Department of Water Resources (DWR), Department of Groundwater Resources (DGR), Department of National Parks, Wildlife and Plant Conservation (DNP);
- Ministry of Industry Thailand: Department of Industrial Works (DIW);
- Ministry of Defence: Armed Forces Development Command (AFDC), Hydrographic Department, Royal Thai Navy (HDRTN);
- Ministry of Transport: Marine Department (MD);
- Ministry of Information and Communication Technology: Thai Meteorological Department (TMD);
- State Enterprises: Metropolitan Waterworks Authority (MWA), Provincial Waterworks Authority (PWA), Industrial Estate Authority of Thailand (IEAT), National Housing Authority (NHA), Electricity Generating Authority of Thailand (EGAT);

- Independent Public Agencies: Department of Drainage and Sewerage, Bangkok Metropolitan Administration (DDS), Office of the National Research Council of Thailand (NRCT).

Generally, people, agencies and organizations at various levels may be involved in water management. An actor is anyone or anything that interacts with the development of water management as shown in Figure 3.7.

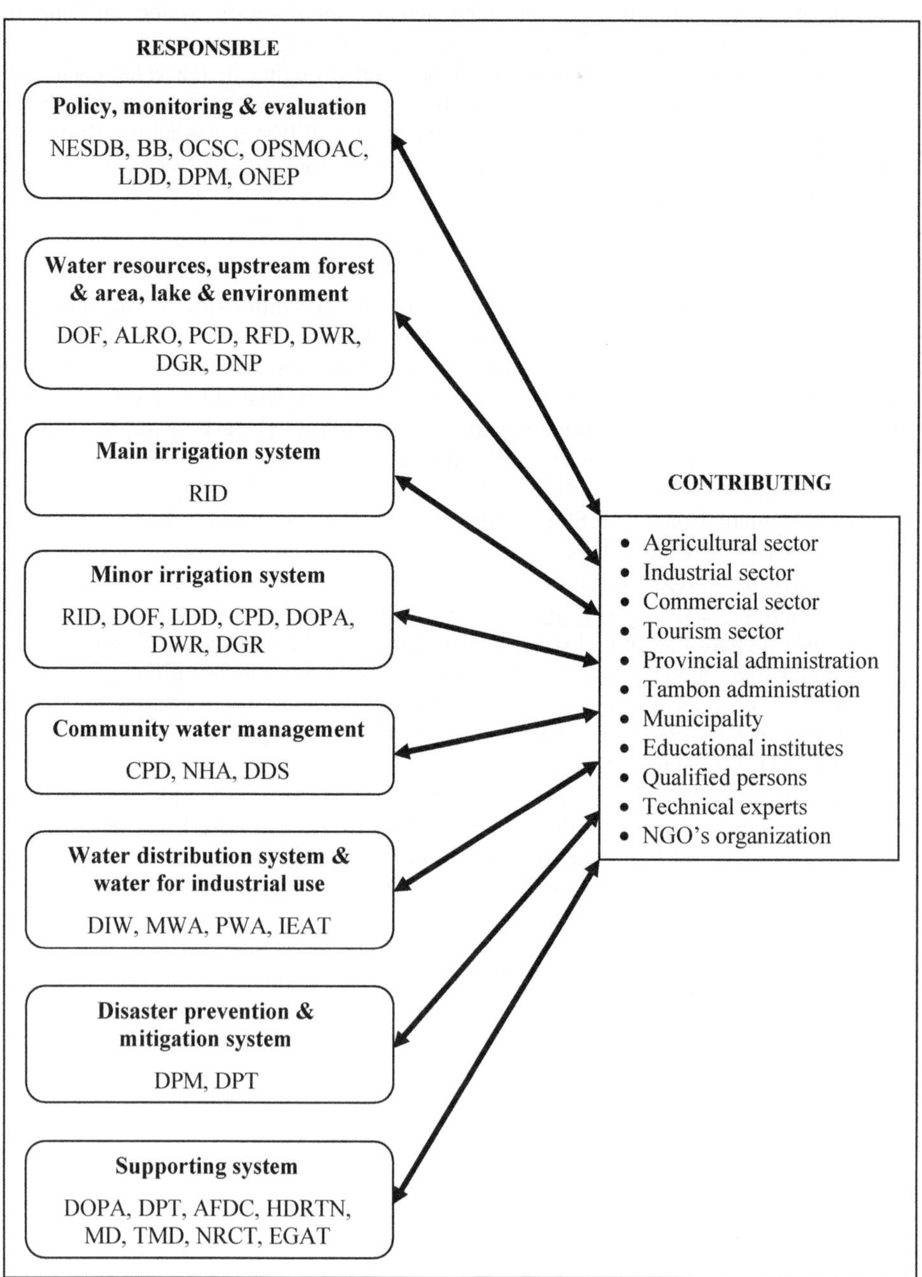

Figure 3.7 Actors interactions in the field of water management in Thailand

In terms of institutional arrangement, the Ministry of Natural Resources and Environment is responsible for the policy and overall planning of natural resources, including water resources. The Ministry of Agriculture and Cooperatives is mainly responsible for the implementation and operation of the infrastructure for the agricultural areas. The water resources management organizations and agencies with water-related mandate are presented in Figure 3.8.

3.5.3 Problems of water resources management in Thailand

Competition for water thus exists between regions, between different sectors, and between upstream and downstream users in river basins and sub-basins. There are also differences in ideas over the most appropriate ways to manage, govern and allocate water in Thailand. The problems of water resources management in Thailand are at present as follows (Suiadee, 2002):

- *government policy*. Government policy did not provide sufficient clear guidelines on water sector management and practices to be adopted. Emphasis has always been placed only on the development of water resources and the provision of water. There were no master plans for water resources management in river basins;
- *institutional problems and constraints*. Problem of fragmentation prevails in water sector management. This makes things complicated and even confusing. Although there are too many executing agencies dealing with water resources, there is no river basin organization to work out water resources development and integrated water management of the river basin;
- *budgeting*. An annual budget is allocated to each agency, based on individual requests by the respective agencies. Such a process is not directly oriented towards problem-solving in those areas and does not address water resources management issues in a holistic manner. Moreover, it usually leads to inefficiencies in implementation;
- *legal framework*. There are several acts concerning water resources but not even one directly relates to water resources management. In order to properly address the increasing problems of more complex requirements of national development, it would be useful to draft a law on water resources management;
- *availability of information*. Because of the relative large number of implementing agencies, information on water resources development scatters all around. This fact makes it difficult to establish plans for efficient water resources development and management. Similarly, it is also difficult to formulate good new projects under such circumstances.

3.5.4 Institutional arrangements for flood protection in Thailand

Manuta et al. (2006) summarised that the present flood disaster management system in Thailand evolved from the Civil Defence Act of 1979 and the Bureaucratic Reform Act in October 2002. The Civil Defence Act prescribes the jurisdiction and responsibility of concerned organizations and a systematic process of disaster management (Asian Disaster Reduction Center, 1999). The bureaucratic reforms in 2002 restructured the institutional arrangements for disaster management in the Kingdom of Thailand (Department of Disaster Prevention and Mitigation, 2005). The key agencies can be classified into two groups: (1) National flood disaster-related agencies, and (2) administrative bodies from national to local level.

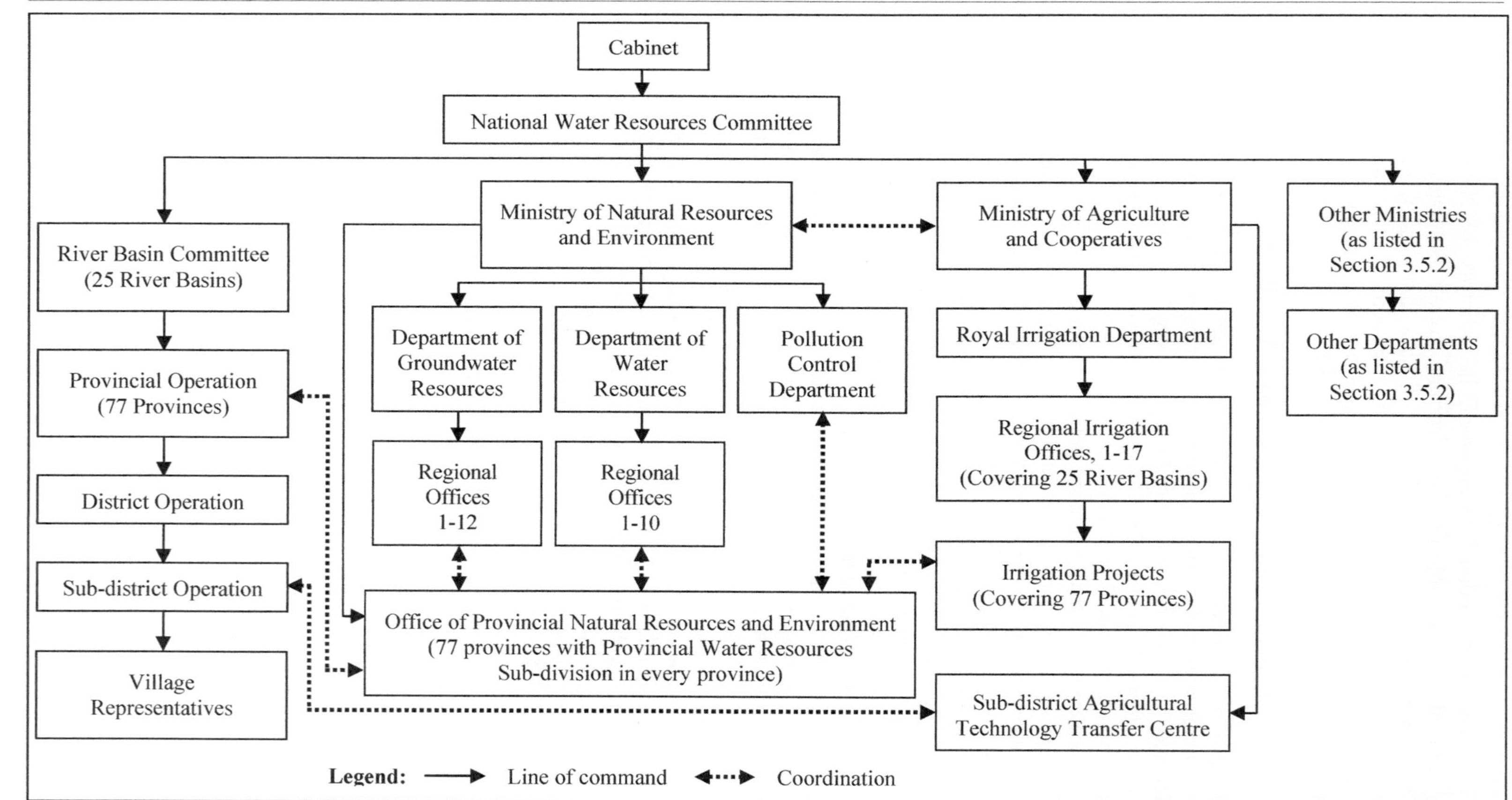

Figure 3.8 Institutional arrangements for water management in Thailand (adapted from Department of Water Resources (2003) and Pattanee (2008))

National flood disaster-related agencies

In October 2002, the government reformed and restructured disaster management in Thailand. The restructuring led to the creation of the Department of Disaster Prevention and Mitigation and assigning of lead responsibilities to this specific department for, more or less, different phases of the disaster cycle, with the aim of ensuring coordination of disaster management nationwide in September 2003 (Table 3.15).

Table 3.15 Roles and responsibilities for disaster management after the October 2002 reforms

Department	Ministry	Responsibility
Disaster Prevention and Mitigation	Interior	Coordinate flood prevention and mitigation plans from different agencies
Water Resources	Natural Resources and Environment	Policy advice on national water policy; plans, coordinate, monitor flood mitigation and national water resource management
All departments	Social Development and Human Security	Carry out rehabilitation and social and economic recovery
Meteorological	Information and Communication Technology	Forecast and early warning systems
Irrigation	Agriculture and Cooperatives	Water provision, storage, maintenance and allocation
Public Works and Town & Country Planning	Interior	Infrastructure rehabilitation and provide development planning guidelines

Sources: Department of Disaster Prevention and Mitigation (2004)
Department of Water Resources (2005a)

As part of the bureaucratic restructuring at the end of 2002, the Department of Water Resources was established as the main agency for formulating management plans, as well as monitoring, coordination and implementation of water resources conservation and rehabilitation (Department of Water Resources, 2005a). It is also the lead agency for flood control and mitigation in the 25 river basins in the country. Mitigation seems part of integrated water management with involvement of the stakeholders (Department of Water Resources, 2005a). It has completed integrated plans for all river basins in 2007 (Department of Water Resources, 2007). Support for the formation of multi-stakeholder bodies known as 'river basin organization' is a key part of its strategy.

The Royal Irrigation Department is one of the more historically powerful agencies for development in Thailand with responsibilities for water provision, storage, maintenance and allocation. Its extensive systems of irrigation canals, gates and pumps are also used as flood protection and drainage facilities in the wet season.

Merging the Departments of Public Works and Town & Country Planning, the new department focuses on one hand the infrastructure rehabilitation after the disaster, and on the other hand, provides development guidelines for urban settlement and environment development. The department proposes policy and development plans in land uses and public works. It adopts zoning to delineate residential areas and categorizes settlement density, and proposes layout of the road network and drainage and flood protection infrastructure.

Several important agencies are involved with flood-related disaster management. As a result of various public pressures following devastating flash floods in Nakhon Sri Thammarat, Southern Thailand in 1988, it concerns:

- the Royal Forest Department imposed a total log ban in natural forests in 1989 (Lakanavichian, 2001);
- the Land Development Department provides zoning of different land use types within the river basin (Economic and Social Commission for Asia and the Pacific, 1999);
- the development of the National Civil Defence Policy Plan in 2002 by the Department of Local Administration provides a master framework and guidelines on developing action plans on disaster management at the provincial and district levels;
- the Electricity Generating Authority of Thailand has the key responsibility in the operation of the multipurpose reservoirs with electric power generation facilities;
- the Marine Department is responsible for navigation in the main rivers and their navigable tributaries, and undertakes channel improvements, which may have flood control impacts;
- the Port Authority of Thailand operates and maintains the port of Bangkok. It maintains and improves the channels of the lower estuary and has a direct interest in flood control works, which would affect the navigability of the rivers (Economic and Social Commission for Asia and the Pacific, 1999).

Administrative bodies from national to local level

With the government reform of 2002, the administration structure for flood management in the country is designated into three levels, which are national, provincial and local levels:

- *at the national level*. The National Civil Defence Committee, which is responsible for formulating policy on disaster management and prevention, is chaired by the Minister of Interior under the authority of the Prime Minister. It likewise, supervises all activities relevant to civil defence and disaster management. The Department of Disaster Prevention and Mitigation is the secretariat of the committee (Department of Disaster Prevention and Mitigation, 2005);
- *at the provincial level*. The Provincial Civil Defence Committee is headed by the Governor of the province and the membership of the committee comprises representatives from various government disaster-related agencies. Under the Civil Defence Act of 1979, the governors are empowered to call different agencies to provide relief in case of major disasters. Flood Mitigation and Prevention Ad Hoc Committees are established in each flood prone province to prepare and plan for the eventual flood disaster. However, some of these committees were only established after a disastrous flood occurred. For example, the Hat Yai Flood Prevention and Mitigation Management Committee, was created after the catastrophic floods in 2000;
- *at the district level*. The District Chief Officer heads the District Civil Defence Committee. Headed by the Mayor, the Municipal Civil Defence Committee comprises the Directors of Bureaus and Divisions of the municipal office. Each municipality is responsible for civil defence and disaster management. The Tambon Administrative Organization prepares the annual budgets for disaster relief and emergency and collaborates with the District Disaster Relief Committee for damage investigation and distribution of compensation in the villages.

4 Impact assessment of land use on flooding conditions

4.1 Beginning

Land is a major resource used to serve vital human needs and wants. However, the actions and practices undertaken on land resource may lead to significant direct and indirect consequences on land conditions, vegetation, and water resources. Given the importance of the impacts of changing patterns of land use on water resources, it is difficult to arrive at a statement that would be universally accepted as land use change is a complex phenomenon that varies greatly from place to place and from time to time. Although it may seem overwhelming in its complexity, it is still essential to quantify its impact on water management systems in order to obtain optimal mitigation and integrated water resources management strategies to cope with present and future risks of extreme flood events.

4.2 Definition of land use, land cover and land use changes

Despite the importance of land resources, there are many different aspects and definitions relating to land use. However, there seems to be some general agreement on land use defined by Clawson and Stewart (1965) that land use is man's activities on land, which are directly related to the land, while Food and Agriculture Organization and United Nations Environment Programme (1998) illustrated that land use is characterized by the arrangements, activities and inputs people undertake in a certain land cover type to produce, change or maintain it. Patterns of land use in different areas have been naturally evolved from different cultures and customs through their common practices. Land use patterns can also be officially regulated by land use planning through zoning and building permits or planning permission laws, or even by private contractual agreements such as restrictive covenants.

Baulies and Szejwach (1997) proposed a clear definition of four basic aspects: land cover, land cover change, land use, and land use change as given below:

- *land cover*, refers to the physical characteristics of the earth's surface. It is defined by the distribution of vegetation, water, desert, ice, and other physical features of the land, including those created solely by human activities such as mining exposures and settlement;
- *land use*, is the intended employment of a land management strategy placed on land cover type by human agents or land managers. Forest, a land cover, may be used for selective logging, for resource harvesting, or for recreation and tourism. Shifts in intent and/or management constitute land use changes;
- *land cover and land use changes*, may be classified into two broad categories: conversion and modification. First, conversion refers to changes from one cover or use type to another. Second, modification, in contrast, involves maintenance of the broad cover or use type in the face of changes in its attributes. Land cover conversion operates through many pathways, the consolidation of which forms specific processes. In turn, these pathways are typically activated by changes in the use of land, or specific operating strategies that are linked to changes in the purpose

of land management. It is important to recognize that many land use/land cover change pathways exist and are differentiated globally, regionally and over time.

Huising (1993) described about land use zone and land use pattern as shown in the details below:

- land use zone refers to a geographical unit or object with a particular land use pattern and a dynamic behaviour, expressed by change in the land use characteristics and by change in the land use zone boundaries. The land use zone provides a geographical basis for the description of land use and land use change at sub-regional level;
- land use pattern describes the land use of a land use zone. The land use pattern denotes the land utilization types and farming systems occurring within a land use zone.

4.3 The influence of land use change on river basin system

Changes in land use associated with urbanization, forestry, agriculture, drainage, or channel modifications can result in changes in a river basin system. There are a number of mechanisms by which land use changes can impact on water storage, which are discussed below.

The process of land use changes has considerable hydrological impact in terms of influencing the nature of runoff and other hydrological characteristics (Ott and Uhlenbrook, 2004; Masih et al., 2011). Land use change (e.g. urban development) within the upper parts of river basins and outside floodplains can increase flood flows, and source controls designed to store and delay runoff are an important means of reducing this problem (Gardiner, 1994; Gardiner, 1998). Changes in vegetation cover and surface slope usually have an impact on total evaporation, interception, infiltration, overland flow and channel flow. In addition, changes in vegetation cover and surface slope may also result in higher erosion levels, which occur quite common (Quiroga et al., 1996).

Brooks et al. (2003) mentioned the effects of land use change on the hydrological processes as follows:

- rainfall is the result of meteorological factors. In relation to effects of land use changes, rainfall can to some extent be controlled by land use (Uhlenbrook, 2007; van der Ent et al., 2010). Land use and associated vegetation alterations can affect the deposition of rainfall by changing interception, at least to some extent. The total interception is the sum of:
    - water stored on vegetative surfaces (including forest litter) at the end of the storm;
    - the evaporation from these surfaces during the storm. In addition, the flow of water into, through, and out of a river basin all can be affected by land use and management activities.
- the type, density and coverage of vegetation on a river basin influence transpiration losses over time (Jewitt, 2005). Differences in transpiration rates among individual vegetation types and vegetation communities can be attributed largely to differences in rooting characteristics, stomatal response, and albedo of plant surfaces. Moreover, the annual transpiration losses are also affected by the length of a plant's growing season;
- changes in vegetation that reduce annual total evaporation will increase streamflow and/or groundwater recharge. On the contrary, increases in annual total evaporation

have the opposite effect (Sun et al., 2008). The best estimates of vegetative effects on actual total evaporation come from water budget analyses and micrometeorological investigations;

- activities that compact or alter the soil surface, soil porosity and permeability, or the vegetative cover can reduce the infiltration capacity of a soil. Exposing a soil to direct raindrop impact also will diminish the openness of the surface soil and reduce infiltration capacities. Land use can also affect infiltration capacities indirectly by altering soil moisture content and other soil characteristics. For instance, soil infiltration is reduced by intense husbandry practices, which are caused by heavy machinery and high stocking densities, the long-term effects being soil degradation and compaction leading to overland flow and erosion;
- upland forested river basins usually are viewed as being important recharge zones for aquifers, because forests in tropical river basins occur in areas with high annual rainfall and are standing often at soils that have high infiltration capacities. It is possible that extensive soil disturbance in a recharge zone could cause groundwater-fed, yearly streams to become dry during seasonal low-flow periods. Such occurrence might be rare, however, and would be significant only where small catchments feed a localised groundwater aquifer and vice-versa. A realistic assessment of land use impacts on the recharge of large, regional groundwater aquifers indicates that little impact would be expected under most conditions. However, there are specific situations such as riparian-wetland systems where changes in vegetative cover can affect groundwater directly. Wetlands are usually low-lying areas connected with the groundwater system. Peat lands may occur as extensive wetlands. Vegetation on wetlands can be forest, shrubs, mosses, grasses, and sedges. Being able to predict the effects of wetland alterations on groundwater (and vice-versa) requires knowledge of the surface-groundwater linkages as well as an understanding of the hydrological processes affected (Wenninger et al., 2004);
- land use activities that change the type or extent of vegetative cover on a river basin will frequently change water yields (López-Moreno et al., 2011) and, in some cases, maximum and minimum streamflow. Changes in vegetative cover occur as a normal part of natural resource management and rural development. Studies conducted throughout the world have demonstrated that annual water yields change when vegetation type or extent is substantially changed on a river basin. In general, changes that reduce total evaporation increase water yields. Water yield usually increases when:
    - forests are clear-cut or thinned;
    - vegetation on a river basin is converted from deep-rooted species to shallow-rooted species;
    - vegetative cover is changed from plant species with high interception capacities to species with lower interception capacities;
    - species with high annual transpiration are replaced by species with low annual transpiration.

The amount of water yield change and/or change in flood runoff generation depends mainly upon the soil and climatic conditions and the percentage of the river basin affected. The type and extent of vegetative cover can influence the amount and timing of streamflow from a river basin. For example, converting from deep-rooted (trees) to shallow-rooted (grass) vegetation can increase water yield in many areas.

McCuen (2003) compared the changes in agricultural land development and changes in urban land cover as detailed below:

- changes in hydrologic response due to natural or human-induced causes can change the storage characteristics of the river basin. Agricultural land development often involves both land clearing, which reduces surface roughness and decreases natural storage, and improvements in the hydraulic conveyance of the drainage ways including additional drainage system. These changes increase the runoff, which decreases the opportunity for infiltration and recharge. Thus, peak runoff rates will increase, times to peak will decrease, and volumes of the surface runoff will increase. With less infiltration and recharge, baseflow rates will most likely decrease. Increases in flow velocities, which can be major cause of flood damage, accompany the increases in runoff rates, with the higher velocities generally increasing the scour rates of channel beds;
- changes in urban land cover reduce both the natural interception and depression storages and the potential for water to infiltrate, especially where the change involves an increase in impervious surfaces. The hydrologic effects of urban land use changes are similar to the effects of agricultural changes. The reduction in both surface storage and roughness increases the runoff rates and decreases times of concentration. Because of the reduced potential for infiltration, the volume of surface runoff also increases as a river basin becomes urbanized. The construction of storm sewers and streets can also decrease the time to peak and increase peak runoff rates.

Pikounis et al. (2003) used the Soil and Water Assessment Tool (SWAT) to simulate the main components of the hydrologic cycle, in order to study the hydrological effects of specific land use changes in a river basin of the river Pinios in Thessaly, Greece. The results of the examination of three land use change scenarios, namely expansion of agricultural land, complete deforestation and expansion of urban areas, showed an increase in discharge during wet months and a decrease during dry periods. The deforestation scenario was the one that resulted in the greatest modification of total monthly runoff.

Quiroga et al. (1996) discovered that the land surface is subject to continuous change due to natural and man-made causes. As the landscape in a river basin is altered in both space and time, the factors that influence the hydrologic response of the river basin also change. The response of a river basin may vary depending on the specific relationship between land structure and land use. A result of the survey of hydrologic effects due to a river basin changes for various land use types is shown in Table 4.1.

Table 4.1 Some river basin changes and their possible hydrologic consequences (adapted from Singh (1989))

River basin changes	Direct runoff volume	Peak flow	Lag time	Flood frequency	Low flow	Sediment yield	Groundwater supply
Agricultural:							
- Contour farming	√	√	√	√		√	
- Terrace farming	√	√	√	√		√	
- Furrow farming	√	√	√	√		√	√
- Tillage operation	√	√	√	√	√	√	
- Drainage	√	√	√	√		√	√
Urban:							
- Imperviousness	√	√	√	√	√	√	√
- Drainage	√	√	√	√		√	
- Population							√
- Tree planting	√	√	√	√	√	√	
- Industrialization							√

Table 4.1 Some River basin changes and their possible hydrologic consequences (adapted from Singh (1989)) (cont'd)

River basin changes	Direct runoff volume	Peak flow	Lag time	Flood frequency	Low flow	Sediment yield	Groundwater supply
Forest:							
- Afforestation	√	√	√	√	√	√	√
- Fires	√		√				
- Drainage	√	√				√	
Highways:							
- Drainage		√	√	√		√	
- Bridges		√	√	√		√	
- Embankments		√	√	√		√	
Mining:							
- Restoration of landscape	√	√	√	√		√	√
- Drainage		√	√			√	
- Mined material disposal							√
Point changes:							
- Dams		√	√	√		√	
- Channel improvement	√	√	√	√		√	

4.4 The impacts on flood regimes

A flood is one of the extremes of streamflow that result from meteorological events and/or other causes (see Section 1.1). In this sense, an understanding of the hydro-climatological controls on floods and their spatial and temporal variations is essential for estimating flood magnitudes and frequencies. However, anthropogenic alterations to the streamflow regime may result in changing hydrological response to flood producing processes. Therefore, efforts to ameliorate the adverse impacts would have to begin with the determination of response of vegetation pattern to alteration of hydrological conditions, in particular for the primary determinant of vegetation composition and structure.

Brooks et al. (2003) identified that flooding concerns are often downstream, far away from upland river basins where vegetative cover is undergoing change. Peak discharges and associated flood stages along major streams and rivers represent the accumulated flows from many river basins of diverse topography, vegetation, soils, and land use. Increases in peak discharges from any headwater river basin can have little effect on downstream peaks because of the routing and desynchronization that normally occur. However, when stormflow volumes are increased from upland river basins, they may not be damped to the extent that peaks are and result in a cumulative effect on downstream volumes and peak discharges. The combination of increased stormflow volumes and increased amounts of sediment deposition in channels can increase the frequency with which streamflow exceeds channel capacity.

Brooks et al. (2003) also stated that changes in land use, particularly changes in forest cover, will affect rather smaller floods with the annual probability of exceedance of 5% to 20% than major floods with a 2% annual probability of exceedance or greater (López-Moreno et al., 2006). Roads and culverts in rural areas, campgrounds, and small upland communities will generally be affected by river basin changes more than large urban centres and agricultural areas along major rivers. Floods of large river basins are affected more by meteorological factors than by land use activities in upland river basins.

Land use activities affect storm response and flooding in the following ways (Brooks et al., 2003):

- removal of vegetation or conversion from plants with high annual transpiration and interception losses to those with low losses can increase stormflow volumes and the magnitude of peak flows. Such practices can also expand the source areas of flow. After a given rainfall event, antecedent soil moisture and water table will tend to be higher; consequently, less storage is available to hold rainfall from the next event, and source areas are expanded;
- activities that reduce the infiltration capacity of soils, such as intensive grazing, road construction, urbanisation and logging, can increase surface runoff. As the proportion of rainfall that results in surface runoff increases, streamflow responds more quickly to rainfall events, resulting in higher peak discharges. Activities that increase infiltration capacities would be expected to have the opposite effect;
- the development of roads, drainage ditches, skid trails and alterations of the stream channel can change the overall conveyance system in a river basin. The effect is usually an increase in peak discharge caused by a shortened travel time of flow to the river basin outlet;
- increased erosion and sedimentation can reduce the capacity of stream channels at both upstream and downstream locations. Flows that would have remained within the stream banks previously can now flood.

The above impacts can have a noticeable effect on stormflow volume, peak magnitude, and timing of the peak for rainfall events that are not extreme in terms of amount and duration. However, it is generally expected that as the amount and duration of rainfall increase, the influence of the soil-plant system on flood generation diminishes (Bruijnzeel, 1990).

For events other than the extreme, stormflow characteristics change in relation to the severity of disturbance of the soil-plant system and to the part of river basin affected. The important factors that would have to be considered in evaluating impacts of land disturbance on stormflow include the:

- extent of change in vegetative cover, particularly as it relates to changes in interception and antecedent soil moisture condition;
- soil moisture storage and hydraulic properties, the presence of water-impeding layers in the soil, and the changes in soil properties;
- mechanisms of streamflow production and the extent of changes in infiltration capacities or the variable source area;
- changes in detention and retention storage associated with channels, ponds and reservoirs;
- changes in the conveyance system of the river basin that affect the time of concentration of flow; roads and skid trails and their orientation with respect to land slope and proximity to stream channels;
- extent of surface erosion, gully erosion, and mass movements (mudslides, landslides, etc.) in relation to detention storage on the river basin and in the conveyance system.

4.5 Simulation of changes in the water balance under changes in land use

Many changes to the natural hydrologic balance occur due to land and water alteration and urbanization by humans. Land use has a significant impact on the hydrologic regime of the river basin. A natural river basin is in equilibrium. The human impact via land use may bring this equilibrium out of balance (DeBarry, 2004).

4.5.1 Conceptualization of the land system

Quiroga et al. (1996) introduced the conceptualization of the land system for hydrologic modelling, in order to determine the distribution and movement of water. An example of a water movement model in which the land component of the hydrologic cycle is conceptualized by use is shown in Figure 4.1. According to this model, water use is distributed among the various types of land use present, and a basic assumption is made in the sense that water does not change use.

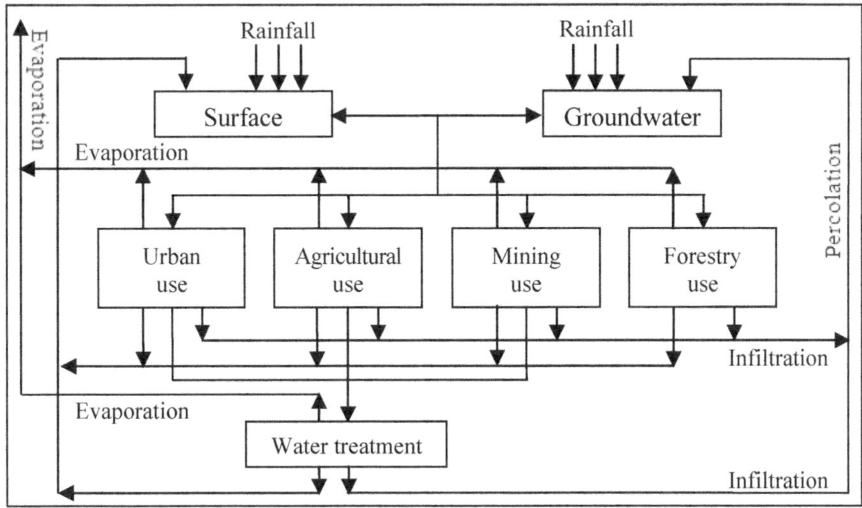

Figure 4.1 Schematic distribution of water according to land use (adapted from Quiroga et al. (1996))

In some cases, especially for ungauged, remote river basins or highly homogeneous river basins, the land system structure within each land use type can be assumed to be uniform. In many other cases, however, especially where there is a significant variability in land use types, it may be required to make the land system structure explicit. Consider the case of Figure 4.2, in which the land system has been divided into three structural subsystems: the vegetation subsystem, the permanent feature subsystem, and the soil subsystem. The permanent feature subsystem refers, in general, to man-made structures, but it can also be extended to include natural geologic controls such as rock formations.

In reality, all these subsystems are interrelated, making the separation of their roles extremely difficult (Uhlenbrook, 2006, 2007). If there are modifications in the characteristics of any one of the subsystems as a result of land use changes in the river basin, the characteristics of at least one of the other subsystems are also likely to change. This, in turn, may result in changes in the distribution and movement of water in all land use types, which means that the entire cycle is altered. For example, deforestation may mean not only the removal of most of the vegetation subsystem, but also the complete transformation of the soil subsystem, and the creation of a permanent feature subsystem. However, the fact that part of the forest is transformed into, say, agricultural areas also implies the creation of water requirements to satisfy agricultural needs which, in turn, may imply changes in the entire hydrologic cycle.

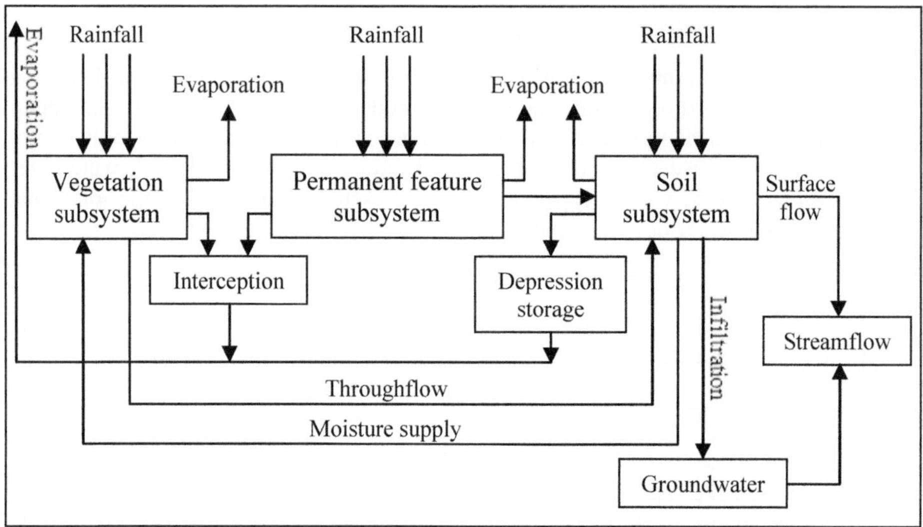

Figure 4.2 Movement of water in the land system (adapted from Singh (1989))

Adequate land use dependent hydrologic simulations depend on how accurately the land use change component itself is modelled.

The model is available according to the period of time in which hydrologic processes are assumed to take place for event and continuous simulation models. The hydrological model is run to simulate the corresponding hydrologic responses. Most models assume unchanging land use and land cover characteristics, which means that the kind of results they produce actually correspond to a river basin in a given state. If the river basin is changing slowly, the use of such an assumption may still be valid. However, if the river basin is changing rapidly, or if, as a result of the change in land use, additional undetected hydrologic processes are triggered or altered, the use of such an assumption may translate into significant levels of uncertainty and errors in simulation results.

4.5.2 Water balance modelling

Event-based models

Event-based streamflow simulation models normally deal with the relationship between rainfall and runoff (Tramblay et al., 2011). Other hydrologic processes are ignored, lumped or very roughly estimated. Event-based models normally compute direct runoff hydrographs (DRH) using a unit hydrograph (UH) (Singh, 2004), or a more physically based approach (Ohara et al., 2011). In the traditional approach, most parameters remain spatially lumped. If data regarding the spatial variability of land use changes is available, the hydrologic modelling of the river basin can be significantly improved by using spatially distributed models (better process representation). Regression analysis could still be used to correlate changes in parameters with spatial indicators of land use change, but their potential is limited as land use patterns rarely follow a nice, clear mathematical function. A better approach is to conceptualize the relationship between land use change and the corresponding hydrologic response at the smallest spatial unit. Clearly, for this approach to be used efficiently, large amounts of geographic data must be handled, and GIS and remote sensing data can be used to assist in this effort (Quiroga et al., 1996).

Continuous simulation models

Quiroga et al. (1996) described the continuous streamflow simulation, which has many more parameters than event-based models. Total evaporation, interception, infiltration, depression storage, sub-surface flow and baseflow are included in more process-based manner. The river basin is divided into sub-basins or other elements (raster cells, Hydrological Response Units (HRUs), etc). Obviously, land use and land cover patterns are often separated along sub-basin boundaries.

Bultot et al. (1990) investigated the impacts of assumed land use changes in the Houille River Basin in Belgium, which were simulated by means of a conceptual hydrological model. The largest difference in impacts appears between river basin-wide coverage by coniferous forests and pastures. Indeed, the mean annual effective total evaporation was maximal for 100% coniferous forests (552 mm/year) and minimal for pastures (477 mm/year), while the mean annual streamflow was minimal for coniferous forests (556 mm/year), and maximal for pastures (631 mm/year). Therefore, under a pasture cover there were 14 low-flow days less and 10 flood days more than under a coniferous forest cover. The discharges associated with extreme flood and low-flow conditions were, however, almost unaffected. In addition, the enlargement of impervious areas clearly modified the streamflow regime in the river basin by inducing more floods as well as longer low-flow stages.

Brooks et al. (2003) presented the modelling of hydrologic effects of river basin modifications. Examples of some continuous simulation models that have a more physical basis are listed in Table 4.2. As can be clearly seen, models have been developed to simulate streamflow response to specific land use practices for a particular region and/or ecosystem. Nevertheless, model capable of predicting hydrologic effects due to changes in vegetative cover or land use practices have also been developed for global application.

Table 4.2 Examples of continuous simulation models that have been used to examine the hydrologic effects of land use changes (adapted from Brooks et al. (2003))

Model	Processes simulated				Flow		Application	Reference
	Rainfall	Snowfall	Infiltration	ET	Surface	Sub-surface		
HSP	√	√	√	√	√	√	Simulates streamflow records for forested, rangeland, agricultural, and urban river basins; engineering applications.	(Hydrocomp Inc., 1976)
PROSPER			√	√	√	√	Simulates water flow through soil-plant-atmosphere system; computes water yield for different soil-plant systems.	(Goldstein et al., 1974)
USDAHL-74	√	√	√	√	√	√	Simulates streamflow records for agricultural river basins; considers zones of infiltration and exfiltration; similar to variable source area approach.	(Holtan et al., 1975)
WBMODEL	√	√	√	√	√	√	Simulates hydrologic changes resulting from river basin management in Colorado subalpine zone; produces year-round water budget.	(Leaf and Brink, 1973)
PHIM	√	√	√	√	√	√	Simulates streamflow records for forested upland and peatland river basins in Upper Midwest; computes effects of timber harvest and peat-mining activities.	(Guertin et al., 1987; Barten and Brooks, 1988)
SWRRB	√		√	√	√		Simulates hydrologic and related processes in ungauged rural river basins; computes effects of river basins management changes on output.	(Arnold et al., 1990)
TAC^D	√	√	√	√	√	√	Simulates all hydrological processes in a fully distributed, but in a conceptual way.	(Ott and Uhlenbrook, 2004; Uhlenbrook et al., 2004)
SWAT	√	√	√	√	√	√	Process-based model for large river basins, suitable for agricultural river basins.	(Arnold et al., 1998; van Griensven et al., 2006; Srinivasan, 2010; Masih et al., 2011)

5 Mathematical modelling

5.1 Initiation

The success of river basin management depends on a robust understanding of the complexity of rainfall-runoff and hydraulic processes. One of the important questions from the hydrological view point, which needs to be clarified in order to respond to the complex nature of the river systems and their hydrological-hydraulic interactions, is how runoff is influenced by the substantial variability of river basin characteristics and human interventions. In this respect, it is undeniable that the use of detailed mathematical models can provide useful tools that allow accurate, robust and reliable management decisions to be made.

In the next section, due attention has been paid to the use of deterministic hydrological and hydraulic models to facilitate the understanding of the physical dynamics between the key interacting components of the hydrological system. The intention is therefore to define the purpose and effectiveness of certain mathematical models in order to carry out detailed flood modelling, to calculate various flood parameters and finally to study and test flood management options in a way which is fair and realistic for both present and future conditions.

5.2 Spatial distribution analysis

Information on land use pattern, i.e. its spatial distribution and change processes, is essential for dealing with the manifold and complex problems of land and water resource utilization and management. Unfortunately, for various reasons, such as a low frequency rate of catastrophic events and the difficulty of predicting their spatial occurrence, active spatial planning and land management systems have not been designated in the past. In this context, it is worth mentioning that understanding of the spatial situation plays a crucial role to ensure its compatibility with the development of landscapes. In addition, the influence of land use characteristics on spatial variability requires not only understanding and explaining past and current land use patterns, a projection of what will happen to future spatial distribution of activities and land uses is an even greater challenge. However, it is not an easy and straightforward process to work towards a better understanding of the dynamics of change in relation to its driving factors and ultimately trace their consequences.

A step in investigating the impact of such change on floods, i.e. identifying critical locations in the face of flood hazards, would also need to be subject to more stringent conditions. To answer the question of what are likely to be the land use change impacts on extreme flows, the finding was in accordance with Schumann and Schultz (2000), the impacts of land use change depend not only on the overall changes in land use types but also on their spatial distribution. In other words, the river basin response will vary on a site-specific basis depending on the land use to be replaced, with an uncertainty on how the land use will be changed. In addressing the problems of conflicting land use that would threat water management from a hydrological perspective, it is useful to investigate the effects of spatial distribution on hydrological response using a spatially (semi-) distributed river basin model with land use dependent parameters derived from concurrent land use data. The changes in flood peaks might appear to be significantly influenced by the spatio-temporal variations of land use. The effects of a distributed spatial accounting for the hydrological response are analysed through a comparison of simulated flood hydrographs provided by different configurations of land use spatial pattern for different sub-basins.

Moreover, the significance of the spatial variations can also be considered adequate for model applications addressed to flood management purposes. It provides the missing basis for taking necessary precautions against flood catastrophes to ensure its compatibility with the prevailing flood situation. As such, an integrated flood management, which is planned to be implemented, can be highly effective, meanwhile the spatial variation should never be oversight as well and needs to be tailored to suit individual situations considering the full range of flood risk.

5.3 Computational models of hydrological systems

To begin with an obvious point, the critical importance of hydrological conditions is placed more firmly within the context of current and reasonably foreseeable uses of land and water resources. From the perspective of hydrological and water resources modelling in the Chi River Basin, a careful attempt would have to be made to identify possible critical hydrologic conditions that may cause failure and conflict of sustainable development of water resources systems.

It is of particular importance to note that the modelling system, which is a choice of a stand-alone constitutive model used to tackle the problems of a hydrological analysis, is never straightforward or simple. It will inevitably need to suit the local situation at any specified spatial-temporal scale with the focus on any element of interest.

The models are usually based on physical properties of elements that define the hydrological processes. For more details, the following aspects are also included:

- hydrological modelling, which is extensively used to estimate design flood hydrographs, is used as an input to the hydraulic model;
- hydraulic model, which is extensively used to route floods through rivers, along floodplains, and flooded areas, is used to calculate local water surface levels and flow velocities of floods of different annual probability of exceedance, and eventually estimate the effects of proposed flood management measures.

A brief description of these broad categories of mathematical models, which are used to represent the various components of river basin water management, is given below with the intention of providing an overview of the model structure and general features.

5.3.1 Classification of hydrological models

As a matter of fact, river basin scale hydrological models must be thoroughly tested to river basins of different sizes and characteristics from different geographic and geologic locations before applying them to evaluate best management practices, and making water resources management strategic decisions.

To address the issue of classification of hydrological models, they are divided broadly into two groups: deterministic models that seek to simulate the physical processes in a river basin involved in the transformation of rainfall to streamflow, whereas stochastic models describe the hydrological time series of the several measured variables such as rainfall, evaporation, and streamflow, involving a probability distribution (Shaw, 1994).

In this section especially deterministic models will be discussed and further classified according to the way account is given to the spatial variability of input and runoff characteristics. The following classification uses a set of criteria presented by Ford and Hamilton (1996):

- *event and continuous*, this distinction applies primarily to models of river basin runoff processes. An event model simulates a single storm and concentrates mainly on the direct runoff, the duration of the storm may range from an hour or less to several days depending on the size of the river basin and the hydroclimatic conditions. A basic problem in modelling of events is the assumption to be made on the initial moisture condition and subsequent rainfall losses. Since an event model is unable to keep the record of soil moisture conditions of the river basin in a continuous manner, it is therefore not useful for process-based modelling. A continuous model operates over an extended period of time, predicting river basin response both during and between rainfall events. Determining discharges and conditions during all periods irrespective of the magnitude of flow: direct runoff periods and periods of no direct runoff. Therefore, the full hydrograph can be reproduced either from measured or synthetic rainfall and potential total evaporation, where based on soil moisture accounting, the dischargeable rainfall is divided in direct runoff, groundwater flow and actual total evaporation;
- *distributed and lumped*, a distributed model is one in which the spatial (geographic) variations of river basin characteristics and processes are considered explicitly, while in a lumped model, these spatial variations are averaged or even ignored;
- *conceptual and empirical or physically based*, this distinction focuses on the knowledge base upon which the mathematical models are built. A conceptual model (grey box models) is built upon a number of conceptual elements with a base of knowledge of the relevant physical, chemical, and biological processes that act on the input to produce the output. On the contrary, an empirical model (black box model) is based on transfer functions, which relate inputs and outputs, without seeking to represent explicitly the process of conversion, while in the physically based models (white box models), full account is given to the physics of the processes;
- *measured-parameters and fitted-parameters*, this distinction is critical in selecting models for application when observations of input and output are not available. A measured-parameter model is one in which model parameters can be determined from system properties, either by direct measurement or by indirect methods that are based upon the measurements. On the contrary, a fitted-parameter model includes parameters that cannot be measured. Instead, the parameters must be found by fitting the model with observed values of the input and output, or estimated indirectly. Strictly, at river basin scale none model exists where all parameters can be observed directly.

The following is a brief explanation about stochastic modelling approach, even though this point will not be considered further. In this respect, it is an undoubted fact that the reality of all hydrological phenomena is stochastic in nature, therefore an attempt is made to generalize a mathematical framework for the nature of stochastic hydrology. Modelling of hydrological stochastic processes can point out the importance of considering the chronological sequences of hydrological events with the aims of attempting to explain the irregularities of occurrence and forecasting the incidence of outstanding important extremes. In particular, accounts for the randomness in the variations in the timing and magnitude of extreme rainfall (Shaw, 1994).

Traditionally, a stochastic model is derived from a time series analysis of the historical record. The stochastic model will be used for the generation of long hypothetical sequences of events with the same statistical properties as the historical record. One technique of generating several synthetic series with identical statistical properties is denoted the Monte Carlo technique. These generated sequences of data can be used in the analyses of design variables and uncertainty (Li, 2006).

In the Chi River Basin, there are several issues, which need to be investigated such as flood analysis, land use changes impacts, damage analysis, etc. In this regard, the deterministic, distributed, and process-based models are obviously attractive to apply and provide a description of flow processes above and below the surface (Figure 5.1).

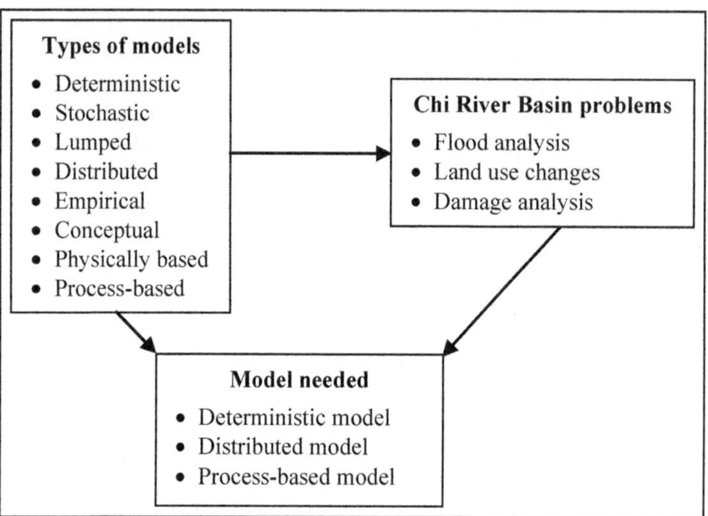

Figure 5.1 The relationship between mathematical models and the Chi River Basin problems

Refer to Figure 5.1, a number of mathematical models have been reviewed by following the sequence of natural processes: rainfall-runoff, water balance, and floods/flooding. Based on the review, the promising models are proposed to be used in this study.

5.3.2 Rainfall-runoff models

As a result of the limitations of hydrological measurement techniques, a rainfall-runoff model may be needed as a means of extrapolation from available measurements in space and time, especially in ungauged river basins where measurements are not available and into the future where measurements are not possible, in order to assess the likely impact of future hydrological change (Beven, 2001). Therefore, the rainfall-runoff model is necessary to be used for converting meteorological inputs (rainfall, total evaporation) into a hydrological output.

Several rainfall-runoff models are widely used, in order to provide a hydrograph showing the variation of volume flow rate (Q) of direct runoff over time at a particular point of interest, usually taken as the river basin outlet, for example, HEC-HMS (US Army Corps of Engineers, 2000), TOPMODEL (Beven, 2001), TAC (Uhlenbrook and Leibundgut, 2002), TOPKAPI (Liu and Todini, 2002), IHACRES (Cunderlik, 2003), MIKE11 Rainfall Runoff (RR) module (DHI Water and Environment, 2007b), SOBEK Rainfall Runoff (RR) module (Delft Hydraulics, 2004), TAC^D (Uhlenbrook et al., 2004), Hydro-BEAM (Smith, 2005), PRMS (Yeung, 2005), SWAT (Neitsch et al., 2005), etc. These hydrological models provide information on the dynamics and the behaviour of the river basin. In the following part, main characteristics of the models will be reviewed. In particular, the main features of HEC-HMS, MIKE11 Rainfall Runoff (RR) module, SOBEK Rainfall Runoff (RR) module, and SWAT will be discussed and outlined briefly.

As an alternative to hydrologic response modelling system, HEC-HMS can also describe an observed historical rainfall event, a frequency-based hypothetical rainfall event, or an event that represents the upper limit of rainfall possible at a given location. It can estimate the volume of runoff, given rainfall amount and properties of the river basin. Moreover, it also includes an automatic calibration package that can estimate certain model parameters and initial conditions, given observations of hydro-meteorological conditions.

With respect to the MIKE11 Rainfall Runoff (RR) module, all hydrologic processes can be modelled and applied for general hydrological analysis, flood forecasting (usually in combination with the hydrodynamic model of MIKE 11), extension of streamflow records, and prediction of low flows.

SOBEK Rainfall Runoff (RR) module can be used to transform rainfall intensities into discharge towards open water. It can also be used in low-lying areas, such as polders. In addition, it provides a wide range of modelling objects, such as unpaved areas, weirs, etc.

The SWAT model computes surface runoff using the SCS curve number method associated with several components that essentially drive the hydrological system. Looking at its capabilities, the model seems to be able to adequately describe rainfall-runoff processes with reasonable accuracy for ungauged river basins.

HEC-HMS

All hydrological models of the rainfall-runoff type involve computational components that reflect river basin storage, losses and the timing of the runoff response. The differences between the hydrological models are the level of detail considered in terms of physical processes and the data input requirements. In this part, the Hydrologic Modelling System (HEC-HMS) will be reviewed (US Army Corps of Engineers, 2000).

Description of type of models in HEC-HMS

Mathematical models, which are included in HEC-HMS are shown in Table 5.1.

Table 5.1 Categorization of mathematical models in HEC-HMS

Category	Description	Remark
Deterministic or stochastic	Deterministic model	All models
Event or continuous	Event models	Most of the models
Distributed or lumped	Lumped models	The ModClark model is an exception
Conceptual or empirical or physically based	Conceptual and empirical models	- The kinematic wave runoff is a conceptual model; - Snyder's unit hydrograph (UH) model is an empirical model
Measured-parameters or fitted-parameters	Measured-parameters and fitted-parameters models	- The baseflow model is a fitted-parameter model; - The Green and Ampt loss model is a measured-parameter model

Source: US Army Corps of Engineers (2000)

Model approaches

The physical representation of a river basin is accomplished with a river basin model. Hydrologic elements are connected in a network system in order to simulate runoff processes. Available elements are sub-basin, reach, junction, reservoir, diversion, source, and sink.

The goal of this study is to determine the area inundated by a storm of a risk to be determined in relation to the land use. Therefore, the model needs to accurately compute and report the peaks, the volumes and the hydrographs of river basin runoff. The HEC-HMS view of hydrological processes as illustrated in Figure 5.2 shows only those components necessary to predict runoff in detail. The other components, i.e. any detailed accounting of movement of water within the soil are neglected or lumped. HEC-HMS includes models of infiltration from the land surface, but it does not model storage and movement of water vertically within the soil layer. It combines the near surface flow and overland flow and models this as direct runoff. It does not include a detailed model of interflow or groundwater flow, instead representing only the combined outflow as baseflow.

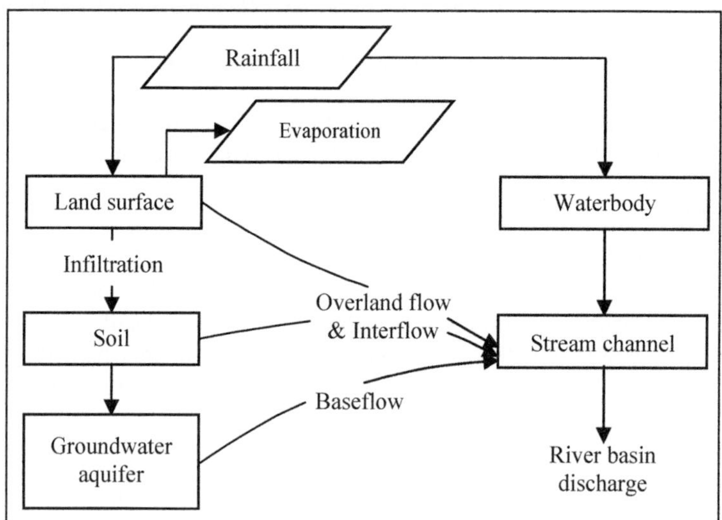

Figure 5.2 Typical HEC-HMS representation of river basin runoff (US Army Corps of Engineers, 2000)

Short-term storage of water throughout a river basin, namely in the soil, on the surface, and in the channels, plays an important role in the transformation of rainfall excess to runoff. The linear reservoir model is a common representation of the effects of this storage, which is based on the water balance principle.

Model capabilities

HEC-HMS is designed to simulate the rainfall-runoff processes of river basin network systems. This includes large river basin water supply and flood hydrology, and small urban or natural river basin runoff. Hydrographs produced by the model are used directly or in conjunction with other software for studies of water availability, urban drainage, flow forecasting, future urbanization impact, reservoir spillway design, flood damage reduction, floodplain regulation, and systems operation. Eventually, this hydrological model is well documented and freely available in public domain.

Different methods for analysing historical rainfall and producing synthetic rainfall are included. Potential total evaporation can be computed using monthly average values.

Different methods are available to simulate infiltration losses. Several methods are included for transforming excess rainfall into surface runoff by unit hydrograph methods. Multiple methods are included for representing baseflow contributions to sub-basin outflow. The linear reservoir method conserves mass by routing infiltrated rainfall to the channels.

A variety of hydrologic routing methods are included for simulating flow in open channels. Channels with trapezoidal, rectangular, triangular, or circular cross sections can be modelled. Water impoundments can also be represented. Lakes are usually described by a user-entered storage-discharge relationship. Reservoirs can be simulated by describing the physical spillway and outlet structures. Pumps can also be included as necessary to simulate interior flood area. Control of the pumps can be linked to water depth in the collection pond and, optionally, the stage in the main channel.

Simulation results can be viewed from the river basin map. Global and element summary tables include information on peak flow and total volume. A time series table and graph are available for elements. Results from multiple elements and multiple simulation runs can also be viewed.

Most parameters for methods included in sub-basin and reach elements can be estimated automatically using optimization trials. Observed discharge must be available for at least one element before optimization can begin. Parameters at any element upstream of the observed flow location can be estimated.

The power and speed of the program make it possible to represent river basins with hundreds of hydrologic elements. Therefore, a Geographic Information System (GIS) can use elevation data and geometric algorithms to perform the same task much more quickly.

Model limitations

The limitations that arise in this program are due to two aspects of the design, which are simplified model formulation and simplified flow representation, as shown below:

- simplifying the model formulation allows the program to complete simulations very quickly while producing accurate and precise results;
- simplifying the flow representation aids in keeping the computation process efficient and reducing duplication of capability in the HEC software suite.

This model is deterministic, which means that the boundary conditions, initial conditions, and parameters of the model are assumed to be known. The program also uses constant parameter values, which means they are assumed to be time stationary. During long periods of time it is possible for parameters describing a river basin to change as the result of the human or other processes at work in the river basin. These parameter trends cannot be included in a simulation. There is a limited capability to break a long simulation into smaller segments and manually change parameters between segments.

This model consists of several modules, i.e. total evaporation and infiltration, and they are uncoupled. The program first computes total evaporation and then computes infiltration. To solve the problem properly the total evaporation and infiltration processes should be simulated simultaneously with the mathematical equations for both processes numerically linked. Errors due to the use of uncoupled models are minimized as much as possible by using a small time interval for calculations.

The river basin model allows each hydrologic element to have only one downstream connection, so it is not possible to split the outflow from an element into two different downstream elements. The diversion element provides a limited capability to remove some of the flow from a stream and divert it to a different location downstream in the network, it will require a separate hydraulic model which can represent such networks. In addition, the design of the process for computing a simulation does not allow for backwater in the stream network.

MIKE 11 Rainfall Runoff (RR) module

Conceptual rainfall-runoff models are widely used in hydrological modelling. One of the well known examples of this type of model is MIKE 11 Rainfall Runoff (RR) module, which will be reviewed in this part (DHI Water and Environment, 2007b).

Description of type of sub-modules in MIKE 11 Rainfall Runoff (RR) module

The MIKE 11 Rainfall Runoff (RR) module is a continuous rainfall-runoff model of the deterministic, lumped, conceptual type. The Rainfall Runoff module of MIKE 11 also includes event-based models based on the SCS and other loss models in combination with unit hydrograph methods.

MIKE 11 Rainfall Runoff (RR) module consists of four different sub-modules that can be used to estimate river basin runoff (DHI Water and Environment, 2007b):

- the continuous simulation (NAM) module, a deterministic, lumped, conceptual rainfall-runoff model with moderate input data requirements, simulating the overland flow, interflow, and baseflow components as a function of the moisture contents in each of three different mutually interrelated storages namely surface, root zone and groundwater storages;
- the Unit Hydrograph Module (UHM), simulates the runoff from single storm events by the use of the unit hydrograph technique (SCS method) includes the different loss models (constant, proportional). It constitutes an alternative to the NAM model for flood simulation in areas where no streamflow records are available or where unit hydrograph techniques have already been well established. The module calculates simultaneously the runoff from several river basins and includes facilities for presentation and extraction of the results. The output from the module is used as lateral inflow to the hydrodynamic module;
- SMAP, a monthly soil moisture accounting model;
- urban, two different model runoff computation concepts are available in the rainfall runoff module for fast urban runoff, i.e. Time/area method and Non-linear reservoir (kinematic wave) method.

Model approaches

MIKE 11 Rainfall Runoff (RR) module can be used either for continuous hydrological modelling over a range of flows or for simulating single events. MIKE 11 Rainfall Runoff (RR) module also includes a facility for calculating mean areal rainfall by e.g. Thiessen weighting, combining daily and N-hourly data.

Using rainfall, potential evaporation, and temperature as input, the model simulates:

- interception as described by the surface storage, retaining the initial rainfall;

- total evaporation as taken from the surface storage at potential rate and subsequently from the root zone at a rate that depends on the soil moisture content;
- infiltration by dividing excess rainfall into infiltration and overland flow, and depending on the soil moisture content in the root zone;
- overland flow as assumed to be proportional to the excess rainfall and linearly varying with the moisture content in the root zone;
- interflow, assumed to be proportional to the moisture content of the surface storage and linearly varying with the moisture content in the root zone;
- overland flow and interflow, routed through two linear reservoirs in series;
- groundwater recharge, by considering the amount of infiltration, which recharges the groundwater, dependent on the soil moisture content in the root zone;
- baseflow, calculated as outflow from the groundwater storage, acting as a linear reservoir. The groundwater storage may be divided in two separate storages, representing a slow and a faster recession;
- capillary flux may be taken into account. This is calculated from the depth to the groundwater table and the relative moisture content in the root zone;
- irrigation and groundwater abstraction can be taken into account.

Model capabilities

MIKE 11 Rainfall Runoff (RR) module can either be applied independently or used to represent one or more contributing river basins that generate lateral inflows to MIKE 11 river network model. This makes it possible to treat a single river basin or a large river basin containing numerous river basins and a complex network of rivers and channels within the same modelling framework. Applications include flood forecasting, flood management, flood frequency analysis, prediction of low flows, etc.

Model limitations

Towards more precise representation of reality, it is true that the model is not supposed to confront the inherent limitations which might conflict with the purposes of any given investigation. Therefore, the discussion will focus particularly on certain weaknesses and can be summarised as (Refsgaard and Knudsen, 1996):

- the NAM module, which forms the part of lumped Rainfall Runoff (RR) module, has disregarded the spatial variation of rainfall and used the river basin average series as input;
- for the simulation of ungauged river basins, a subjective evaluation of river basin characteristics is undertaken for estimation of the appropriate model parameters, while distributed models directly use the available information on the spatial variation of topography, soil, vegetation types and their characteristics for model set up and estimation of appropriate model parameters.

SOBEK Rainfall Runoff (RR) module

In many studies, there is still a need for a hydrological model that is more versatile than most of the current rainfall-runoff models. Many of the commonly used rainfall-runoff models involve runoff generation and flood hydrograph routing algorithms. In this section, SOBEK Rainfall Runoff (RR) module will be reviewed (Delft Hydraulics, 2004).

Brief description of SOBEK-Rural/Urban

SOBEK-Rural/Urban is a widely physically based model. It is an integrated software package for river, urban or rural management. Seven program modules, such as water flow (FLOW), Rainfall Runoff (RR), Water Quality (WQ), Real-Time Control (RTC), Sediment Transport (ST), Morphology (MOR), and Salt Intrusion (SI) modules, work together to give a comprehensive overview of water systems.

Model approaches

SOBEK-Rural RR module

The rainfall-runoff process of rural areas and various types of unpaved areas can be modelled, taking into account land use, the unsaturated zone, groundwater, capillary rise and the interaction with water levels in open channels.

SOBEK-Urban RR module

SOBEK-Urban RR module simulates the rainfall-runoff process for various types of paved and unpaved areas. It does not matter whether the urban drainage system consists of open channels and sewer pipes, storage tanks and reservoirs. Even street flow can be modelled.

Model capabilities

SOBEK-Rural RR module

The model capabilities of SOBEK-Rural RR module can be described as follows:
- it can model rainfall-runoff and other hydrological processes in rural and urban areas in a simplified way;
- river basin areas can easily be modelled in a lumped or distributed manner, with no restriction to the number of sub-basins;
- river basin areas can be modelled in any detail using land elevation curves, soil characteristics, land cultivation, drainage characteristics, etc.;
- the model distinguishes between various rainfall-runoff processes such as surface runoff, sub-soil drainage and storage in saturated and unsaturated areas, taking into account crop total evaporation and capillary rise;
- it can use separate storm events or long time series of meteorological data for statistical analysis;
- the user can input any rainfall patterns or use historical data, and model any number of rainfall gauges taking into account the spatial variation;
- it can model both flood events and dry spells.

SOBEK-Urban RR module

Considering the SOBEK-Urban RR module, the modelling capabilities of this module are as follows:
- it can model dry weather flows and rainfall-runoff processes for various types of paved areas, such as streets, roofs and parking lots;
- the model can be extended with possibilities for unpaved areas and groundwater using the fully integrated link with the Rainfall Runoff module of the SOBEK-Rural product line;

- urban areas can be easily modelled in a lumped or detailed manner with no limit to the number of areas;
- the user can input time and spatially varied rainfall patterns or use historical data of storm events and long time series with or without dry periods.

Model limitations

Besides its capabilities, there are also limitations with this model and all main items can be listed as follows:
- it needs some unusual input data, such as area per crop (m^2, ha or km^2), drainage resistance values for each layer (day), etc.;
- the calculations sometimes take longer, i.e. a few minutes up to one hour;
- it is neither freeware nor open source so it is not possible to modify the model structure in order to fit the operational requirements.

SWAT

The Soil and Water Assessment Tool (SWAT) rainfall-runoff model, which is a successful modelling effort of rainfall-runoff transformation and fully developed long-term continuous simulation model in particular in large river basins dominated by agriculture, will be reviewed (Neitsch et al., 2005).

Brief description of SWAT

The SWAT model is a process-based model that simulates basically all hydrologic processes in a river basin. SWAT model can be used as a semi-distributed model by partitioning the river basin into sub-basins and HRUs to account for differences in physical processes such as soils, land use, crop, topography, etc. In addition, SWAT is also a continuous model, i.e. a long-term yield model. The model is not designed to simulate detailed single-event flood routing.

Model approaches

For modelling purposes, a river basin may be partitioned into a number of sub-basins. The use of sub-basins in a simulation is particularly beneficial when different areas of the sub-basin are dominated by land uses or soils dissimilar enough in properties to impact hydrology. By partitioning the river basin into sub-basins, the user is able to reference different areas and to distribute the input parameters (rainfall, temperature, etc.) of the river basin to one another spatially.

Input information for each sub-basin is grouped or organized into the following categories: climate, hydrologic response units (HRUs), ponds/wetlands, groundwater, and the main channel/reach/draining the sub-basin.

In principle, SWAT uses Hydrological Response Units (HRUs) as the basis for its modelling to describe spatial heterogeneity in land cover, topography and soil types within a sub-basin. The HRUs are defined as lumped land areas within the sub-basin that are comprised of unique land cover, soil, topography and management combinations. The model estimates relevant hydrologic components such as total evaporation, surface runoff, soil moisture, and groundwater recharge and flow at each of the HRUs (Kangsheng, 2005). Since the model maintains a continuous water balance, a distributed Soil Conservation Services (SCS) curve number is therefore used for the computation of overland flow runoff. As a result, excess surface runoff not lost to other functions makes its way to the channels where it is routed downstream. The amount of input required for SWAT simulations depends on the purpose of the simulations, the minimal input requirements are topography, land use, soil, and hydro-meteorological data.

The hydrologic process simulated by SWAT is based on the water balance equation given for the soil routine as follow (Figure 5.3 and Equation 5.1):

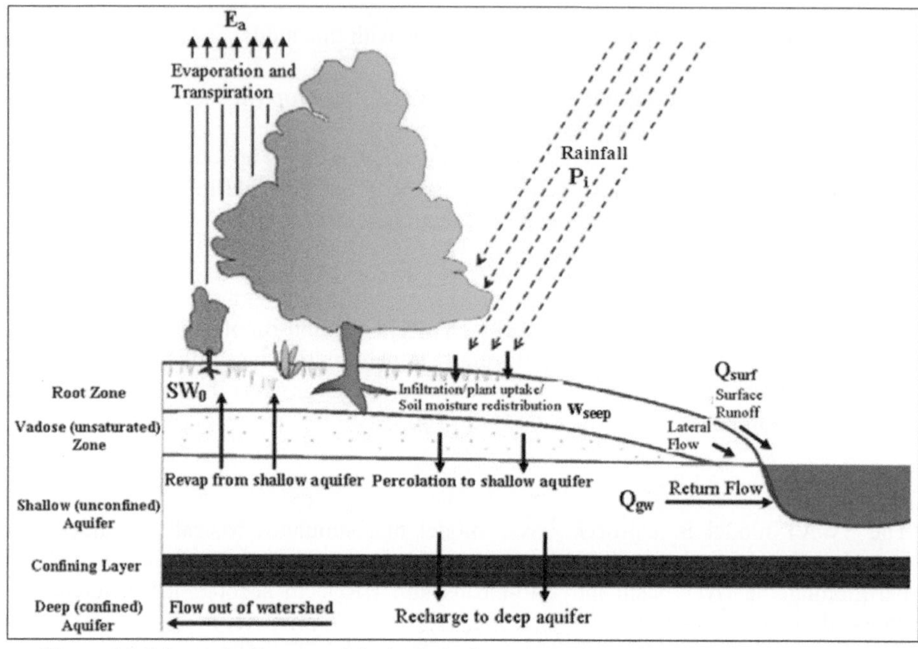

Figure 5.3 Schematic diagram of the hydrologic cycle (adapted from Neitsch et al. (2005))

$$SW_t = SW_0 + \sum_{i=1}^{t} \left(P_i - Q_{surf} - E_a - w_{seep} - Q_{gw}\right) \tag{5.1}$$

where:

SW_t = final soil water content (mm)
SW_0 = initial soil water content on day i (mm)
t = time (day)
P_i = amount of rainfall on day i (mm)
Q_{surf} = amount of surface runoff on day i (mm)
E_a = amount of actual total evaporation on day i (mm)
w_{seep} = amount of water entering the vadose zone from the soil profile on day i (mm)
Q_{gw} = amount of groundwater flow on day i (mm)

The sub-division of the river basin enables the model to reflect differences in the hydrological processes for various crops and soils. Runoff is predicted separately for each HRU and routed to obtain the total runoff for the river basin as shown in Figure 5.4.

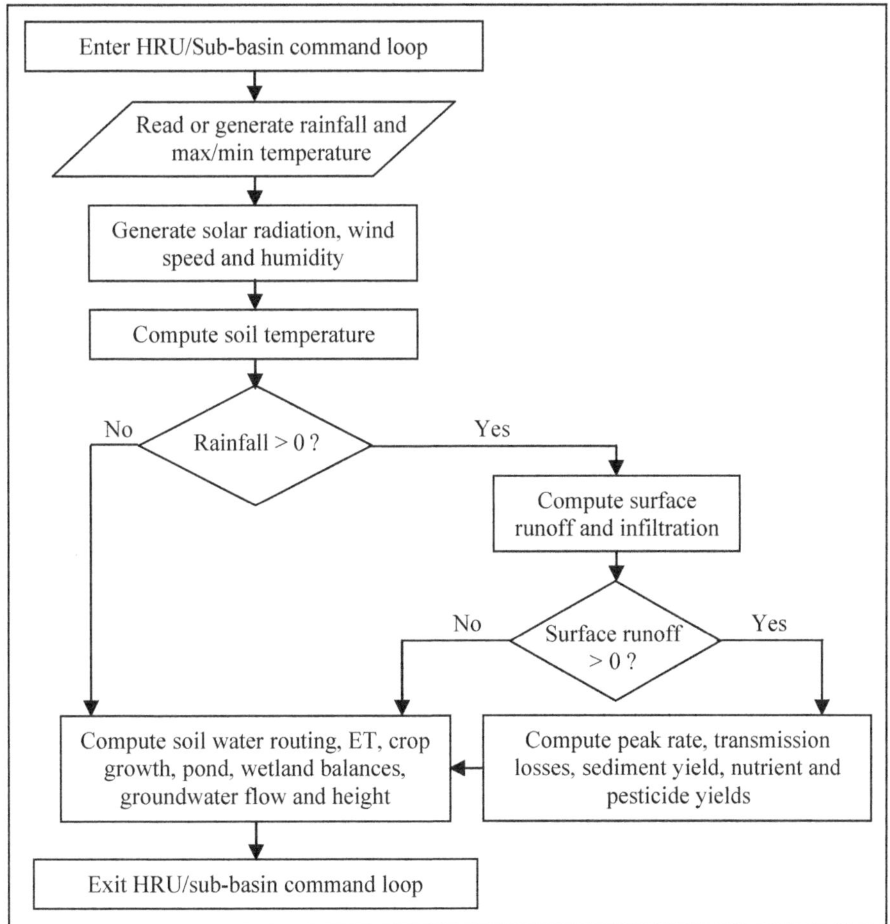

Figure 5.4 The general sequence of processes used by SWAT to model the land phase of the hydrologic cycle (Neitsch et al., 2005)

Model capabilities

SWAT is computationally efficient due to the fact that its simulation in case of very large river basins or a variety of management strategies can be performed without excessive investment of time or money. It is easy to use, good GIS-linked interface and a lot of experiences worldwide, etc (Gassman et al., 2010). Moreover, the model can also enable the users to study long-term impacts that correspond to a number of the problems currently addressed, i.e. the impact on downstream waterbodies.

In this study, applications of the SWAT model are compiled and discussed in order to evaluate the effectiveness of the Best Management Practices (BMP) and the implementation of which can help to reduce the damaging effects of storm water runoff on waterbodies and landscape. In addition, SWAT is a public domain model with two main benefits, which are:

- the impacts on water movement for river basins with no monitoring data (e.g. streamflow data) can be modelled;

- the relative impact of alternative input data (e.g. changes in management practices, climate, vegetation, etc.) on water quantity or other variables of interest can be quantified.

Model limitations

To illustrate this issue, there are some limitations of the SWAT model informing its weak points as presented below.

- limitations in simulating detailed event based flood and sediment routing;
- require extensive data inputs that are not readily available, e.g. soil data with sufficient spatial detail;
- there are too many parameters that need to be estimated, leading to model uncertainty;
- the effects of lumping parameters arbitrarily into HRUs, for example, distributed parameters such as hydraulic conductivity distribution could not be represented;
- it fails to show the interaction between the HRUs as they are not internally linked within the landscape but are all routed individually to the sub-basin outlet. Therefore, the impact of management of an upslope HRU on a downslope HRU cannot be assessed (Arnold et al., 2010);
- it does not actually simulate the groundwater dynamics since its groundwater component is oversimplified (Luo and Sophocleous, 2011).

5.3.3 Models for water balance simulation

The movement of water through the sequence of the atmosphere, vegetation, and soil is an important process. Understanding the water balance in relation to river basin characteristics provides insight into the influences of several different land use scenarios in relation to any seasonal water yield.

There are many water balance simulation models commonly used worldwide that represent the changes in physical processes associated with different land use changes. For instance, GAWSER/GRIFFS (Cunderlik, 2003), HBV (Swedish Meteorological and Hydrological Institute, 2006), LASCAM (Viney, 2003), IBIS (Twine et al., 2004), TAC^D (Uhlenbrook et al., 2004), PRMS (Yeung, 2005), SWAT (Neitsch et al., 2005), WBM (Stephens et al., 2005), DRCM (Bari and Smettem, 2006), WaSiM-ETH (Wagner et al., 2006), etc. These hydrological models have been developed to describe the land phase of the hydrological cycle in space and time.

In this section, HBV and SWAT models, which have equal or better modelling efficiencies than other aforementioned models, will only be reviewed for their capabilities to observe the behaviour of water balance components for different land use concepts in the Chi River Basin. These hydrological models are well documented, freely available in public domain, and can model basically all hydrologic processes in a river basin. Moreover, it is fairly low data requirements for HBV model, and there is a comprehensive model structure which provides a user-friendly GUI (Graphical User Interface) in the SWAT model. In the end, one of them will be selected for this research work. A brief description of both models is given below.

HBV

Another water balance model at river basin scale, the Hydrologiska Byråns Vattenbalansavdelning (HBV) model, which is a deterministic mathematical model of the hydrological processes in a river basin used to simulate the runoff properties, will also be reviewed in this study (Swedish Meteorological and Hydrological Institute, 2006).

Brief description of HBV

The HBV model is a conceptual, event-based, and continuous model, which includes conceptual numerical descriptions of hydrological processes at the river basin scale. It can be used as a semi-distributed model by dividing the river basin into sub-basins.

Model approaches

The HBV model is designed to run on a daily time step (shorter time steps are available as an option) and to simulate runoff in river basins of various sizes (Cunderlik, 2003). The model structure and details of the HBV model are shown in Table 5.2.

Table 5.2 Sub-model description of the HBV model (Seibert, 2005)

Sub-model	Input data	Output data
Soil routine	potential total evaporation, rainfall	actual total evaporation, soil moisture, groundwater recharge
Response function	groundwater recharge, potential total evaporation	runoff, groundwater level
Routing routine	runoff	simulated runoff

Seibert (2005) mentioned that in the response function, the model of a single linear reservoir is a simple description of a river basin where the runoff is supposed to be proportional to the water storage.

Model capabilities

The HBV model is an Integrated Hydrological Modelling System (IHMS). It can be linked with real time weather information and forecast systems. The HBV model assists management, enhance safety, and optimize productions with several applications as shown below:

- flood warning, streamflow and volume forecasting for assessment of flood risks, and development of flood risk maps;
- hydropower, short-term inflow forecasts for operational hydropower planning at transmission centres and volume forecasts of up to a year for seasonal reservoir planning;
- pre-feasibility studies, quality control of water stage and discharge records, extension of historical records and groundwater simulations;
- irrigation, determination of total evaporation and forecasting inflow to reservoirs and storage basins to aid regulation of irrigation schemes;
- dam safety, design flood computations including reservoir management strategies;
- climate change, studies of the effect of changing climate conditions on runoff patterns, soil moisture, groundwater change and total evaporation.

The flexible structure of the HBV model allows the user to make necessarily sub-divisions with respect to different climate zones, land use, density of the hydro-meteorological network, etc.

Model limitations

As with all models, the HBV model has also its own limitations and it cannot possibly explain all aspects of natural phenomena. The list below details some of the potential model shortcomings (Uhlenbrook et al., 2004).

- the model parameters are often not physically based or clearly related to river basin properties;
- there is an existence of consequent uncertainty in predictive capability, i.e. different parameter sets give similar good results during a calibration period, but their predictions may differ when simulating runoff in the future (Seibert, 1997). As a result, this would lead to difficulty in applying the model to ungauged river basins;
- there is also a high degree of uncertainty in model structure as a result of incomplete understanding and simplified descriptions of modelled processes as compared to reality.

SWAT

In this section, the Soil and Water Assessment Tool (SWAT) will be described in relation to water balance studies.

SWAT requires specific information about weather, soil properties, topography, vegetation, and land management practices occurring in the river basin. The physical processes associated with water movement, crop growth, etc. are directly modelled by SWAT (see Section 5.3.2 for details).

Detailed reviews and comparison of SWAT with other models

A detailed review of SWAT model applications, which is the result of various studies that compared SWAT performance and robustness with other hydrological models, has been summarized by Gassman et al. (2007). On the basis of those literature reviews, some of the important ones will be briefly presented.

Borah and Bera (2003) and (2004) compared SWAT with several river basin scale models. The 2003 study reported that all models such as the Dynamic Watershed Simulation Model (DWSM) (Borah et al., 2004), Hydrologic Simulation Program - Fortran (HSPF) model (Bicknell et al., 1997), SWAT, and other models have a hydrologic routine applicable to river basin scale. They concluded that SWAT is a promising model for long-term continuous simulations in predominantly agricultural river basins. The 2004 study showed that SWAT and HSPF could predict yearly flow volumes with adequate monthly predictions except for months with extreme storm events and unusual hydrologic conditions. Moreover, SWAT was poor in simulating daily extreme flow events while DWSM could reasonably predict distributed flow hydrographs at small time intervals. Van Liew et al. (2003) compared the streamflow predictions of SWAT and HSPF on eight nested agricultural sub-basins within the Little Washita River Basin in southwestern Oklahoma. The conclusion was that SWAT is more reliable than HSPF in terms of streamflow predictions of different climatic conditions and hence may be better for investigating the long-term impacts of climate variability on surface water resources. Saleh and Du (2004) found during the calibration and verification periods at five sites in the upper North Bosque River Basin in Texas that the average daily flow simulated by SWAT was closer than HSPF. Singh et al. (2005) revealed that SWAT flow predictions were somewhat better than HSPF estimates for the 5,568 km^2 Iroquois River Basin in eastern Illinois and western Indiana, primarily due to the fact that SWAT could provide better simulation of low flows. El-

Nasr et al. (2005) stated that both SWAT and MIKE SHE models (Refsgaard and Storm, 1995) could simulate the hydrology of Jeker River Basin, Belgium in an acceptable way. Srinivasan et al. (2005) reported that SWAT provided more accurate estimated flow than the Soil Moisture Distribution and Routing (SMDR) model (Cornell University, 2003) for 39.5 ha FD-36 experimental sub-basin in east central Pennsylvania. It was also found that SWAT could provide more accurate calculations on a seasonal basis.

5.3.4 Models for flood simulation

Simulation of flood events that may occur in the Chi River Basin can contribute to minimizing the loss of life and damages to property, and it is essential for flood risk management. In this section, 1D/2D SOBEK (Delft Hydraulics, 2004) and MIKE FLOOD (DHI Water and Environment, 2007a) will be reviewed.

1D/2D SOBEK

There is a growing need for integral water solutions to prevent more loss from the possible effects of floods. With the application of an integrated model called 1D/2D SOBEK for riverine flood simulations, the impact of the integral simulation on hydraulics, flood damage, and flood risk can be analysed to obtain detailed information for flood planning and response.

Brief description of 1D/2D SOBEK

The 1D/2D SOBEK is a physically based hydraulic model. In addition, 1D/2D SOBEK includes a simple distributed hydrological model, which takes into account spatial variation in inputs, outputs, and parameters. In general, the river basin area, which is divided into a number of elements and runoff volumes, is first calculated separately for each element.

Model approaches

Basic equations

The flow in one dimension is described by two equations: the continuity and the momentum equations. The continuity equation in one dimension is expressed as:

$$\frac{\partial A_f}{\partial t} + \frac{\partial Q}{\partial x} = q_{lat} \tag{5.2}$$

where:

A_f = wetted area (m^2)
Q = discharge (m^3/s)
q_{lat} = lateral discharge per unit length (m^2/s)
t = time (s)
x = distance in x-direction (m)

The flow in two dimensions is described by three equations: the continuity equation, the momentum equation for the x-direction, and the momentum equation for the y-direction. The continuity equation for two dimensions is presented by the following equation:

$$\frac{\partial \zeta}{\partial t}+\frac{\partial (uh)}{\partial x}+\frac{\partial (vh)}{\partial y}=0 \tag{5.3}$$

where:

u	= velocity in x-direction (m/s)
v	= velocity in y-direction (m/s)
ζ	= water level above the plane of reference (the same for 1D and 2D) (m)
h	= total water height above the 2D bottom; $\zeta + d$ (m)
d	= depth below plane of reference (m)
y	= distance in y-direction (m)

The momentum equation in one dimension is defined as:

$$\frac{\partial Q}{\partial t}+\frac{\partial}{\partial x}\left(\frac{Q^2}{A_f}\right)+gA_f\frac{\partial h_1}{\partial x}+\frac{gQ|Q|}{C^2RA_f}-W_f\frac{\tau_{wi}}{\rho_w}=0 \tag{5.4}$$

where:

g	= gravity acceleration = 9.81 (m/s^2)
h_1	= water level with respect to the reference level (m)
C	= Chezy coefficient (m$^{1/2}$/s)
R	= hydraulic radius (m)
W_f	= flow width (m)
τ_{wi}	= wind stress shear (N/m^2)
ρ_w	= water density = 1,000 (kg/m^3) at 4 °C

For two dimensional flow, two momentum equations are calculated, together with the 2D continuity equation. The momentum equations are identified by the following equations:

- for x-direction:

$$\frac{\partial u}{\partial t}+u\frac{\partial u}{\partial x}+v\frac{\partial u}{\partial y}+g\frac{\partial \zeta}{\partial x}+g\frac{u|V|}{C^2h}+au|u|=0 \tag{5.5}$$

- for y-direction:

$$\frac{\partial v}{\partial t}+u\frac{\partial v}{\partial x}+v\frac{\partial v}{\partial y}+g\frac{\partial \zeta}{\partial y}+g\frac{v|V|}{C^2h}+av|v|=0 \tag{5.6}$$

where:

V	= velocity; $V=\sqrt{u^2+v^2}$ (m/s)
a	= wall friction coefficient (1/m)

Coupling of 1D and 2D numerical models

Water movement in the stream channel in 1D/2D SOBEK is described by a finite difference approximation, based upon a staggered grid approach as shown in Figure 5.5. The interaction between the 1D and 2D solvers is determined by Equation 5.7. The 1D and the 2D schematisations are combined into a shared continuity equation at the grid points where water levels are defined as illustrated in Figure 5.5 (Frank et al., 2001).

$$\frac{dV'_{i,j}(\zeta)}{dt}+\Delta y\left[(uh)_{i,j}-(uh)_{i-1,j}\right]+\Delta x\left[(vh)_{i,j}-(vh)_{i,j-1}\right]+\sum_{l=1}^{L(i,j)}Q_{k_l}=0 \qquad (5.7)$$

where:

V' = combined 1D/2D volume (m^3)
Δx = 2D grid size in x (or i) direction (m)
Δy = 2D grid size in y (or j) direction (m)
Qk_l = 1D discharge flowing out of control volume through link k_l (m^3/s)
$L(i,j)$ = number of 1D branches connected to 2D nodal point i,j
i,j,k,l = integer numbers for 2D nodal point and 1D channel numbering

The momentum equation is applied at grid points where discharges (1D) or velocities (2D) have been defined. The velocities are eliminated by substitution of the momentum equation into the continuity equation. This results in a system of equations at water level points, with all water levels as unknowns at the new time level.

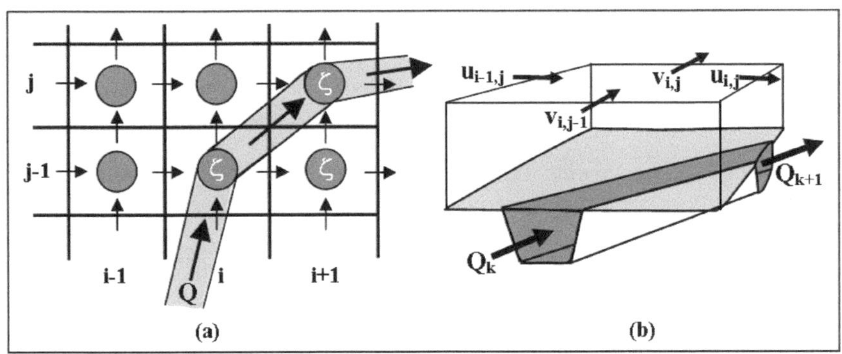

Figure 5.5 Schematisation of the hydraulic model: a) combined 1D/2D staggered grid; and b) combined continuity equation for 1D/2D computations (adapted from Frank et al. (2001))

Model capabilities

The model 1D/2D SOBEK, which offers a fully integrated approach, couples rural and river systems enabling users to model an entire river basin area as one easy-to-manage network. It is necessary to investigate insights to some of the technical features, which enhance its internal suitability for the Chi River Basin. The following are highlights of important aspects of 1D/2D SOBEK capabilities:

- the model has the ability to dynamically simulate complex behaviour of overland flow and its interaction with flow impediments over an initially dry land. Moreover, it can also handle flooding and drying processes on flat or hilly terrain, i.e. there is no artificial water loss while the area of the 2D grid is dry (Dhondia and Stelling, 2002);

- it allows nesting of 2D domains in order to provide more refinement at desired locations;
- the 1D and 2D domains, where they overlap, are automatically coupled and solved simultaneously. As a result, the model allows interactions between the two regimes to be dynamically modelled;
- using a robust numerical scheme, the model is stable enough to handle very complex hydraulic behaviour;
- it is designed to simulate both super-critical and sub-critical flow regimes, including their transitions (Verwey, 2001);
- hydraulic structures can be incorporated into the model to create a realistic picture of how the system behaves in extreme scenarios;
- it works with a GIS-based, user-friendly interface to deliver fast (and accurate) data input and output processing;
- results are displayed as maps, graphs, tables, and animations that allow users to analyse and communicate ideas.

Model limitations

Models, which are used to trace natural phenomena, are inherently inexact and may be imprecise in real life. Despite this weakness, models are still widely in use as general powerful tools to represent natural processes. As with many modelling techniques, there are some known limitations of 1D/2D SOBEK listed below (Kazi Emran Bashar, 2005).

- the model needs many input data sets to make accurate predictions;
- it is unable to define depth-dependent roughness descriptions in the vegetation layer;
- the method used to simulate the role of embankments and their overtopping from the 1D channel to the 2D grid are set to assume highest/lowest level of embankments. Therefore, water will always overtop on both sides of the embankment, it is not possible to clarify whether the water spills towards left or right side;
- the method of overflowing 1D channel with the option assuming no embankments would result in neglecting part of the embankment in the 1D cross-section above ground level;
- the model is mainly suited to the conditions where the channel width is smaller than the grid size of 2D domain.

MIKE FLOOD

MIKE FLOOD is a dynamically coupled one-dimensional (MIKE 11) and two-dimensional (MIKE 21) flood modelling system. It is a state of the art tool, which is targeted specifically towards modelling of floods.

Brief description of MIKE FLOOD

MIKE FLOOD enables the features of both a one-dimensional and a two-dimensional model to be utilized for 1D river hydraulic system and 2D floodplain, respectively. It is especially relevant to floodplain analysis due to:

- comprehensive hydraulic structures package;
- GIS integration for spatial and temporal analysis;
- supercritical flow solutions;
- dam and embankment failure analysis tools.

Model approaches

The fully integrated software system, MIKE FLOOD, consists of MIKE 11, MIKE 21, and a coupling model that links MIKE 11 and MIKE 21. This hydraulic model is based on dynamic linking between the 1D numerical modelling system MIKE 11 for representation of the channel conveyance and the 2D floodplain modelling system MIKE 21 for representation of the 2D dynamic effects associated with out of bank flows. There are three different types of links between MIKE 11 and MIKE 21. A description of each of these link types is described below:

- *standard link*. It is used for dynamic exchange internally in both directions between a detailed MIKE 21 cell/element to the end of a MIKE 11 network, to connect an internal structure (with an extent of more than a grid cell) or feature inside a MIKE 21 cell/element;
- *lateral link*. The link allows flow through the MIKE 11 domain into the MIKE 21 domain, and vice-versa. Flow through this link, which is useful for simulating overflow from a river channel onto a floodplain, is calculated using a structure equation and the water levels in MIKE 11 and MIKE 21;
- *structure link*. It takes the flow terms from a structure in MIKE 11 and inserts them directly into the momentum equations of MIKE 21, which is useful for simulating structures within a MIKE 21 model. The link consists of a 3 point MIKE 11 branch (upstream cross section, structure, and downstream cross section), the flow terms of which are applied to a MIKE 21 cell or group of cells.

Model capabilities

A combined modelling system can utilize the most desirable features, meanwhile minimizing the bad features, i.e. avoiding the limitations of resolution and accuracy commonly encountered when using MIKE 11 or MIKE 21 separately. Within the standard MIKE FLOOD, computational formulations enable many model applications to be improved through its use, including: floodplain applications, storm surge studies, dam break, hydraulic design of structures, and broad scale estuarine applications.

In addition to this fully dynamic description, MIKE FLOOD can be applied on numerous applications with the help of the following special features:

- momentum preservation through 1D/2D links;
- lateral links enabling simulation of overbank areas in detailed 2D with the wider model domain and main river channels in 1D;
- ability to nest 2D grids at finer resolutions to obtain more detail in specific areas;
- ability to model a comprehensive range of hydraulic structures;
- GIS integration for spatial and temporal analysis associated with ease of automated model development and result processing and presentation;
- links possible along any alignment in MIKE 21 (not just horizontally or vertically);
- a graphical user interface standard for data input and output as well as data preparation, presentation, and analysis;
- an on-line help system, user manual, and technical reference documentation.

Model limitations

Though MIKE FLOOD has a number of advantages, there are also some limitations. For instance, the model is often a compromise between model resolutions (grid/mesh size) and computational time required for a simulation, as well as two sub-models (MIKE 11 and MIKE 21) would have to be maintained instead of one. Another limitation is the data requirements since they are often not available, in particular in the developing world.

5.3.5 Evaluation on model selection

The selection of a specific model is always a difficult task. In addition to the capability of the model, it would also equally depend on the capability of the model user and data availability for selected case. This section attempts to compare the selected models introduced in the previous sections. In brief, the mathematical model selection for hydraulic, hydrological and water resources modelling in the Chi River Basin can be classified as follows.

Rainfall-runoff model

Considering choices of models, which have been discussed in Section 5.3.2, HEC-HMS for rainfall-runoff simulation, reservoir routing is not applicable in case of the Chi River Basin, as the reservoirs are operated with controlled outflow. Moreover, reservoir outlet structures and dam break are not incorporated in the model, as a result the model cannot be internally computed.

To get insights in the model selection for rainfall-runoff simulation, it can be found that there are some of the limitations for the MIKE 11 RR module and SOBEK RR module, which are shown as follows.

- *MIKE 11 RR module*:
    - needs to purchase multiple modules to take full advantage of the system;
    - significant data needed to set-up;
    - technical support likely needed;
    - annual subscriptions.
- *SOBEK RR module*:
    - needs some special input parameters;
    - it is not possible to extract, adapt or develop its module as it is too complex;
    - the software is commercial.

Model choice

Out of the selected rainfall-runoff models, SWAT seems to be the best choice from various angles and will be adopted for use. The reasons for choosing this model are that it is an open source model, i.e. it is free, editable and can be adapted to match the particular needs. The model is fast gaining popularity among users as it has been tested successfully under different geographical and climatic conditions, through the continuous improvements done by various groups around the world.

Water balance model

A comparison still has to be done for the water balance calculations using HBV model or SWAT model. Based on a brief comparison and evaluation, finally, one of them has been selected for further research work.

In order to select the water balance model, the model parameters have been taken into consideration. In case of SWAT, which has a comprehensive structure, is able to model basically all hydrologic processes in a river basin. It offers the ability to simulate the complete runoff and to observe the behaviour of water balance components for different land use concepts, which is particularly important in case of water resources management. Unfortunately, the number of parameter of SWAT is much larger than HBV, which may cause a problem in data availability. While considerable attention is then given to HBV according to its advantages such as simplicity, ease of application, and small amount of required input data. However, its shortcomings are nevertheless important and evident, i.e. the ability to model outflow only and no possibility to model changes in the river basin area. For this reason, HBV model is not applied in this study.

Model choice

For the final model selection context, the water balance computation has been done by SWAT as it is found to be a promising model for continuous simulations with a high level of spatial detail in a predominantly agricultural area like the Chi River Basin. Given the aforementioned reviews of the SWAT model application literature, SWAT is capable of simulating hydrologic responses with reasonable accuracy as it is a flexible and robust tool that can be used to simulate a variety of river basin problems.

Flood model

According to the flood model selection, it can be noted that 1D/2D SOBEK and MIKE FLOOD are appropriate models to be used to simulate floods in this study in order to assess damages in the Chi River Basin. Moreover, the model will also be used to assess and improve the effectiveness of current and proposed flood management practices in the Chi River Basin based on the available data of historical flood events.

In this regard, it is found that 1D/2D SOBEK and MIKE FLOOD are comparable since both have the capability to be used for the Chi River Basin study. Among the known advantages that 1D/2D SOBEK can provide include (Kazi Emran Bashar, 2005):

- its ability to deal with complex and irregular channel cross sectional geometries, as well as vegetated floodplains;
- variable roughness parameters in the 2D floodplain domain can be assigned and varied in different parts of the system depending on vegetation pattern for the resistance to composite overland flow;
- provision of an integrated tool suite for mapping flood dynamics.

On the other hand, MIKE FLOOD, which enables both 1D and 2D engines to be performed easily, can also be used to simulate the fully integrated flow dynamics between rivers and surrounding floodplain areas. It ensures modelling flexibility through the use of fine model resolution near critical floodplain features in 2D detail, while other areas can be modelled in 1D, including features such as hydraulic structures with sub-grid scale that are accurately represented in the model.

Model choice

In evaluating hydraulic modelling software packages and making comparisons, it is of interest to note that software architecture is an ever-changing field due to the fact that what was applied in the past may not capture current features and future problems. Therefore, the flood modelling package chosen in this predominantly rural context is at the heart of most of the algorithms described here in order to bring the model behaviour closer to the actual physical behaviour.

Based on a comparative analysis, the model chosen for riverine flood simulations is the 1D/2D SOBEK package, even though there are some specific limitations that are inherent in the modelling approach itself. It is found that 1D/2D SOBEK demonstrates its ability as a robust, reliable, and accurate tool. Moreover, it also offers greater flexibility and features. Another advantage is that 1D/2D SOBEK has been developed by Deltares. When discussion was needed, it was more practical to contact the experts in Delft.

Selection results

With the understanding that no single model is superior to the other for every situation, it is therefore essential to use a suite of mathematical models in order to meet necessary conditions for acceptance from the perspective of its intended uses, rather than reliance on a single model.

In accordance with the model categories outlined above, the components within the modelling suite make it possible for a wide range of applications to complex flows and water related processes in any aspect of river system. As a result of the evaluation on model selection, the relationship between selected mathematical models and physical aspects in the Chi River Basin is established as shown in Figure 5.6.

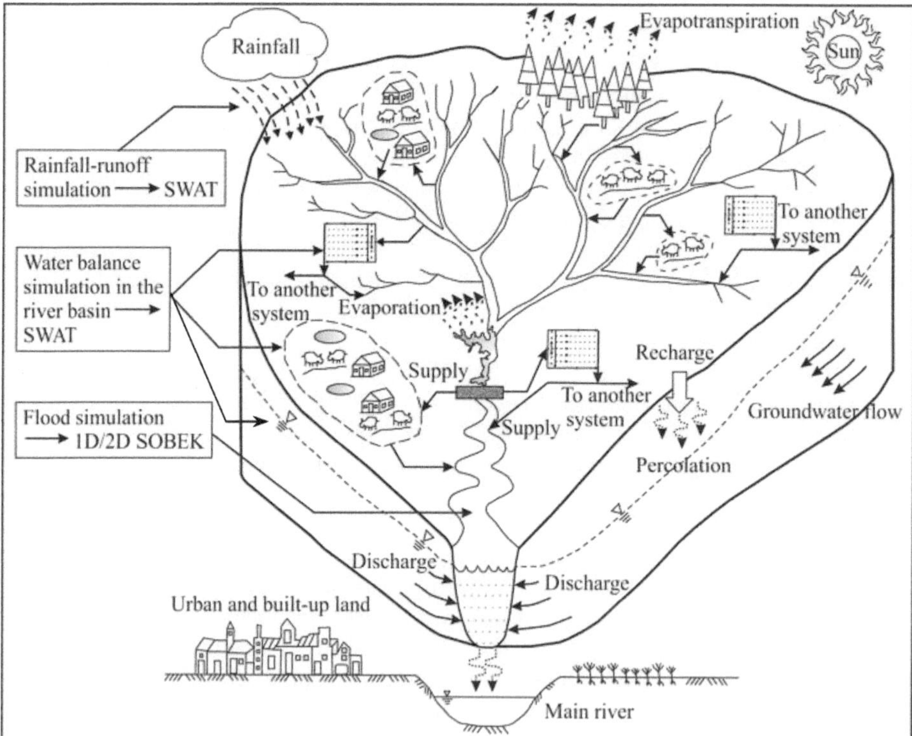

Figure 5.6 Relationship between hydrological and hydraulic models and physical aspects in the Chi River Basin

5.4 Application of the model coupling

After answering the question, which models would have to be coupled, another question arises, i.e. how to couple these models to satisfy the special needs. The Delft-FEWS was selected, which is an integrated package of tools that permits the integration of large data sets, specialised modules to process the data, and the integration of external models to be built and run in the same graphical environment with a GIS-based interface (Werner et al., 2004; Werner and Heynert, 2006). Principally, Delft-FEWS system can carry out the following tasks:

- import of hydro-meteorological data, i.e. rainfall, discharge, and water level time series, from external data bases;
- data transformation in order to prepare the required inputs for hydrological and hydraulic models;
- execution of arbitrary hydrological and hydraulic models;
- visualisation of results on maps and time series.

The mechanism of Delft-FEWS is a sequence of sequential steps, where the input and output requirements are defined at each modelling step. Delft-FEWS provides a platform that connects data streams and multi-model ensembles, i.e. a user interface around the rainfall-runoff model based upon the SWAT algorithm and the hydraulic modelling systems 1D/2D SOBEK, through the use of defined interfaces and standards in data exchange for data handling and through open interface architecture, respectively. Each model component is used as plug-in in Delft-FEWS, and is conducted with a stand-alone application in which it can directly communicate/interact with the interface of Delft-FEWS. In this application, a workflow is defined to conduct the SWAT model for the calculation of runoff flowing into the Chi River and its tributaries. Subsequently, the workflow engine adapts its execution and assigns the overland flow computation activity to 1D/2D SOBEK hydraulic model with the generated runoff from SWAT for its input. A flow chart showing the modelling process is given in Figure 5.7.

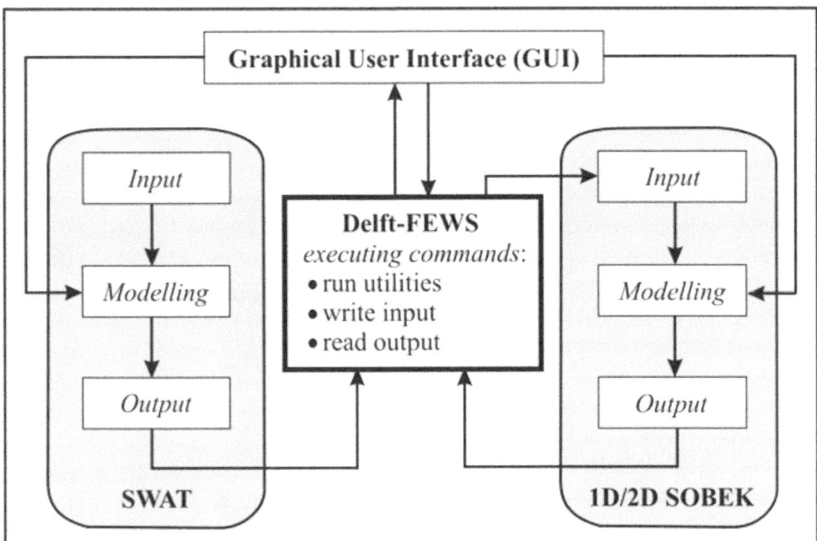

Figure 5.7 Flow chart of the interface system coupling of the models SWAT and 1D/2D SOBEK

The structure of Delft-FEWS is composed of a set of workflows, each of them consists of one or more sub-workflows or activities to provide a sequence, which can identify the configured modules to be run. The workflow engine that defines its internal behaviour is executed independently and sequentially with different time steps, e.g. daily time step (SWAT) and half-hourly time step (1D/2D SOBEK). Figure 5.8 illustrates a series of modelling tasks/workflows facilitated by Delft-FEWS, which was applied to the Chi River Basin. All workflow components are configured and conducted with external tools, i.e. XML-editors and scripts that support the exchange of data between different sub-systems. Moreover, with the help of MATLAB script (The MathWorks Inc., 2008), a workflow execution starts to conduct time series data processing and simulation modelling activities, and import into Delft-FEWS as an ensemble (Figure 5.9). Some screenshots of the graphical user interface of Delft-FEWS, which refer to the case of the Chi River Basin, are shown in Figure 5.10 to 5.14. Depending on the complexity of the workflow, large data volumes may need to be processed in long running workflows. Moreover, any modification or alteration made will be effective upon a rerun of the sub-workflow in order to get the most up-to-date results.

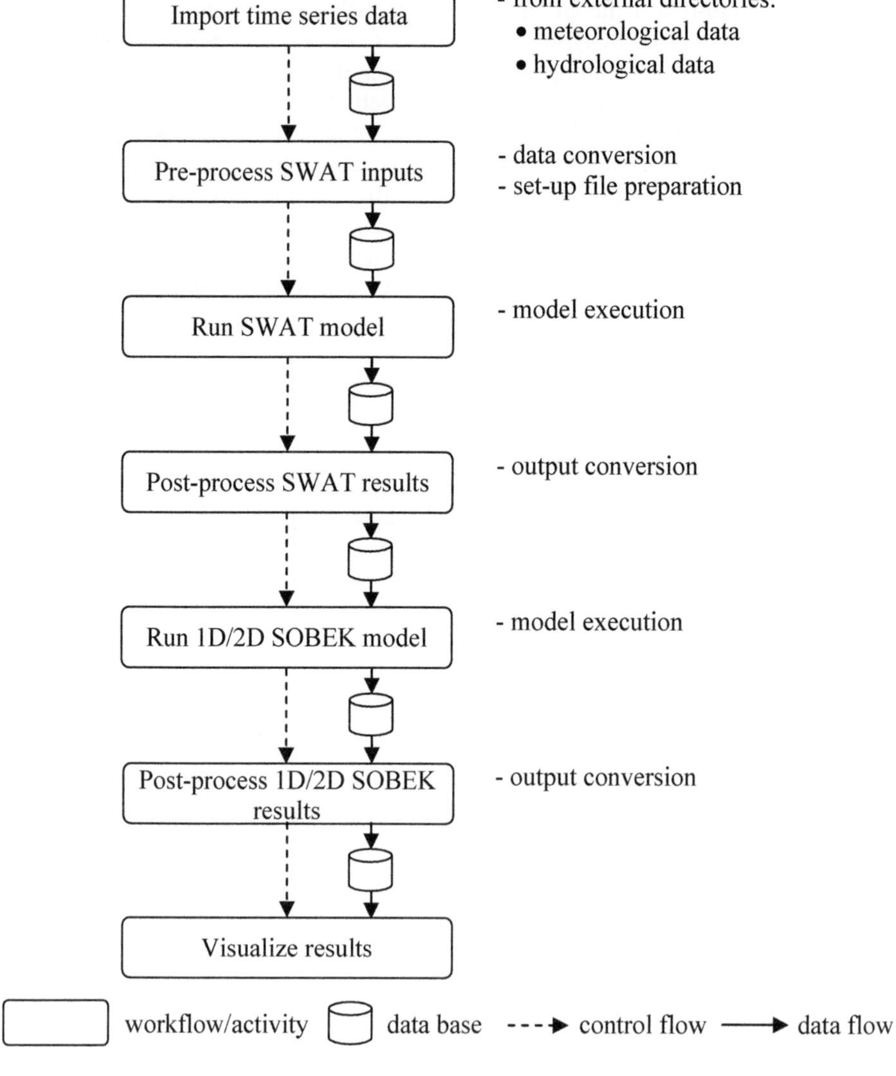

Figure 5.8 Delft-FEWS workflow for the coupling of the models SWAT and 1D/2D SOBEK

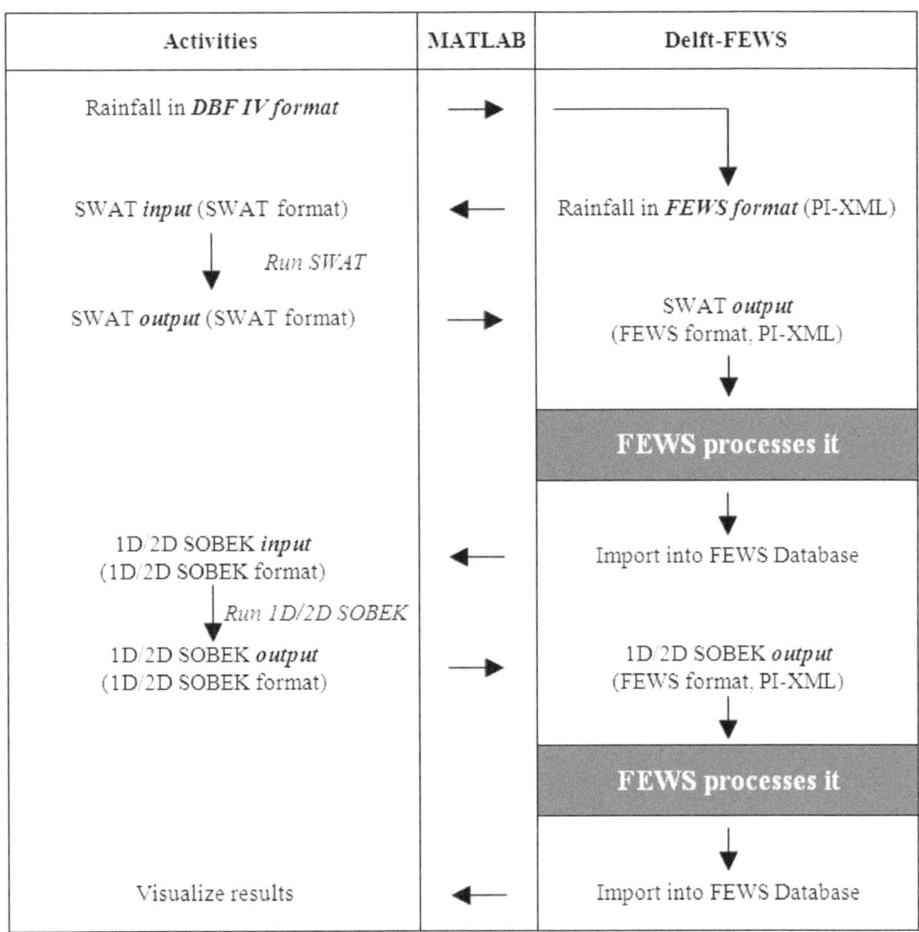

Figure 5.9 Schematic overview of Delft-FEWS configuration

For clarity, a connection to the modelling systems is made up through a software adapter, by which each adapter contains a General Adapter and a Model Adapter. The General Adapter is used to deliver data between a wrapped external model and Delft-FEWS data base via the Delft-FEWS Published Interface (PI) XML format (Deltares, 2010). While the Model Adapter, which is associated with an individual model, is used to convert the data in PI-XML format into the native format for each model, as well as executing the models. After model execution, the Model Adapter will collect and convert the model outputs into the PI-XML format, which will be assimilated by the General Adapter (note: the detailed configuration guide will be prepared in the subsequent stage). It is noticed that all essential functional elements are embedded in the model adapter, while Delft-FEWS only manages time series and model data sets, as well as model states.

It is important to mention that the application of the model coupling, which provides an open architecture framework, allows a more flexible strategy to be chosen. Through its modular structure, and support of open and published interfaces, the system can be easily extended to include additional data formats and models. Numerous models have been included in the system as described previously. With this open approach, changes to the models and data used have no impact on how the system is used operationally, thus it helps to reduce the organisational impact of adaptation, and allows easy integration of advances in data provision and modelling (Werner and Whitfield, 2007). This seems to be a logical strategy in developing an integrated modelling framework for flood management, which provides flexibility in changing underlying models and reflects a growing realisation that no single model concept is suitable for all cases.

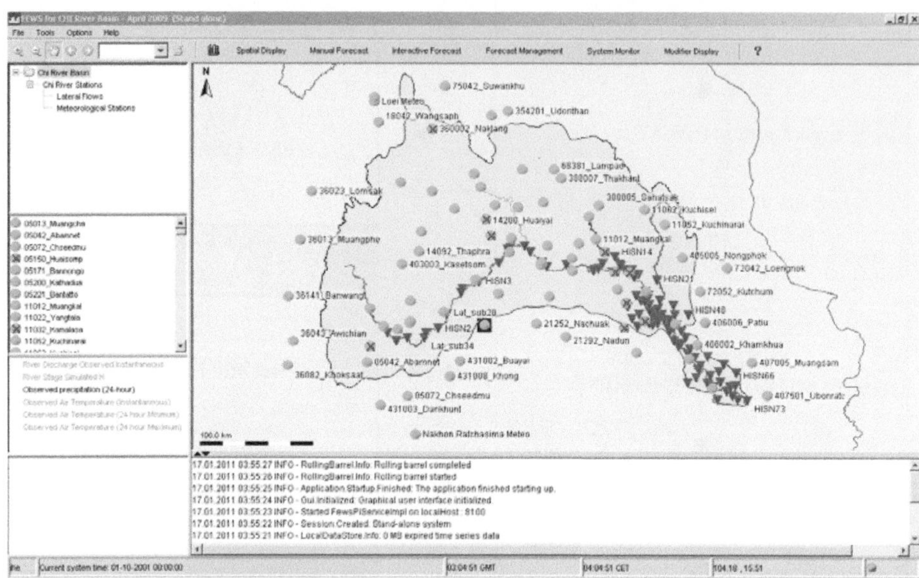

Figure 5.10 Graphical User Interface of Delft-FEWS, as configured for the Chi River Basin

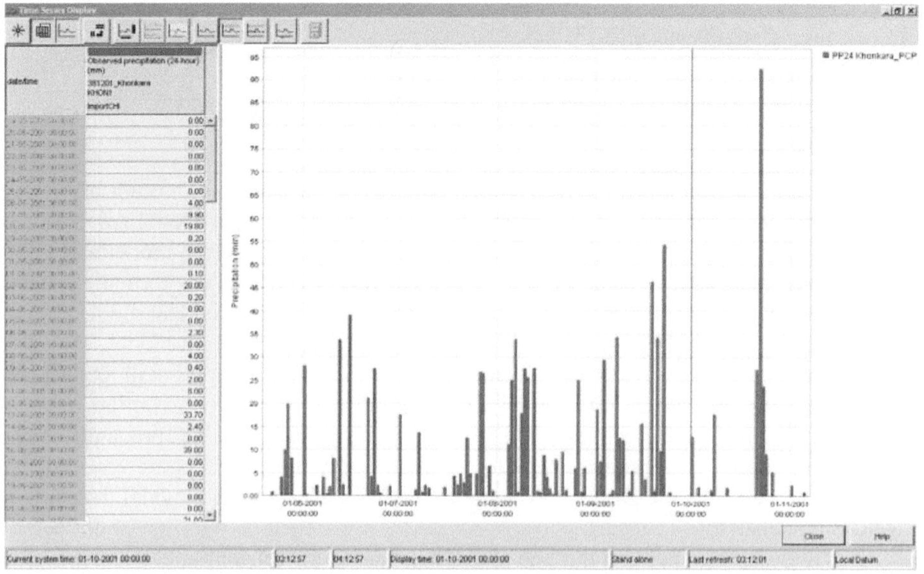

Figure 5.11 Rainfall time series display and editor in the edit mode

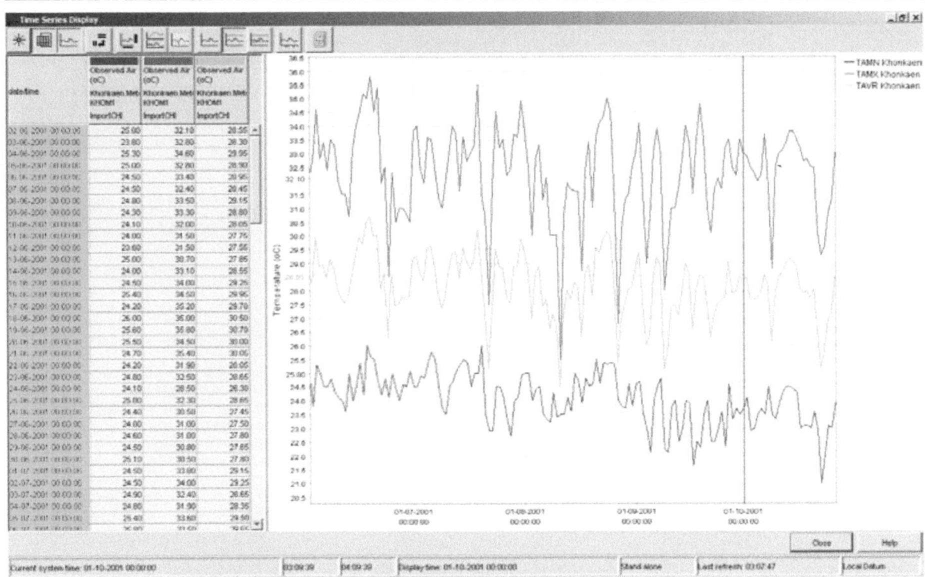

Figure 5.12 Temperature time series display and editor in the Edit mode, showing multiple time series in one subplot

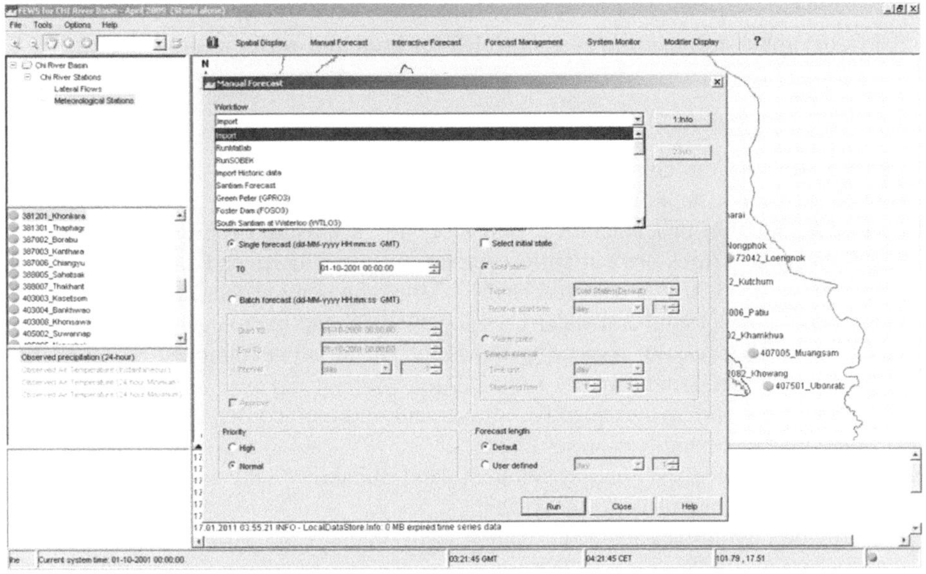

Figure 5.13 Workflow drop-down list for the selection of a pre-configured workflow

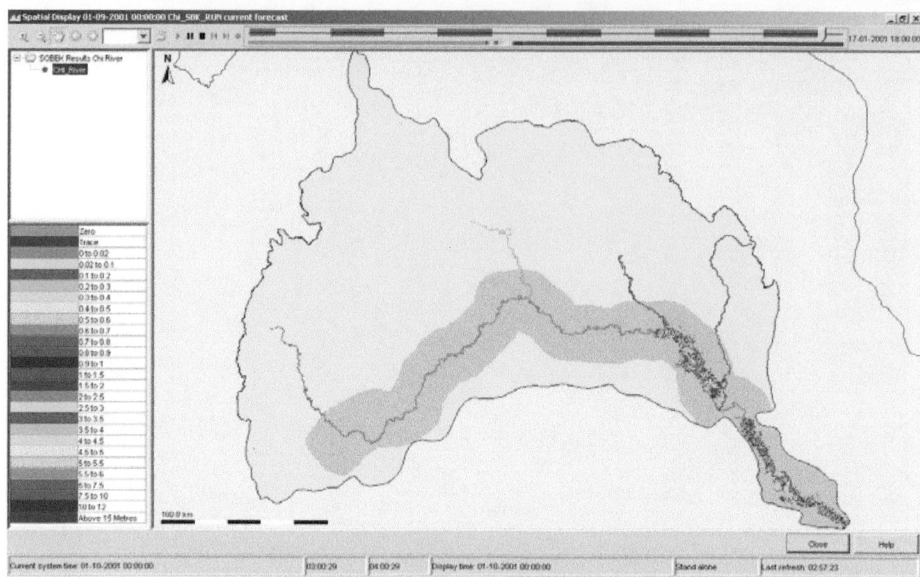

Figure 5.14 Grid Display window in Delft-FEWS environment, showing the flooding in the Chi River Basin (note that shaded area indicates floodplain marsh)

5.5 Geographic information system (GIS) for land and water management

5.5.1 GIS applications for planning and management

Indeed, attempts through human endeavours to pursue various aspects of life rely heavily on the consumption of land and water resources in order to secure their basic necessities, wealth, and amenities. Likewise, this acquisition reinforces the fact that development pressures are currently encroaching the Chi River Basin, impacting the land and water demand, and accelerating stress on natural systems. Consequently, inappropriate use of land and water causes geological and hydrological hazard risks. In this sense, the stresses of increasing utilization of the aforementioned resources have created a rising need for GIS to:

- understand and avoid negative direct and indirect consequences from human activities involving land-water linkages;
- further enhance the interpretation of land suitability and water availability for development and management plans;
- provide possibilities for better spatial analysis that responsive to the needs of the concerned people.

Detailed information of GIS is designed to facilitate integration and analysis of spatially and geographically referenced information, e.g., identified data of each location. Generally, it is designed to effectively collect, process, manipulate, analyse, store, and display spatial information, which provides up-to-date, reliable, accurate, and suitable alternatives for efficient management of large and complex data bases. In hydrology and land-water resources management, GIS is increasingly becoming an important tool to combine the spatial data, i.e. topography, land use and soil maps, and hydrologic and hydraulic variables.

In the aftermath of the flood disaster, the ability of GIS has been recognized as a powerful means to analyse and identify the areas at high risk of flooding. With GIS, different scenarios, i.e. critical what-if questions, can be analysed to help in better land and water resource management, timely decisions, and thus effective management and support of flood disaster relief.

5.5.2 Land use and water management systems zoning

Land use and water management systems zoning is a knowledge-based procedure used to stop or reverse the accelerating degradation, and to mitigate the adverse effects of destructive use of land and water resources. Local people usually have their own traditional strategies to manage land and water resources at variable and interacting spatial and temporal levels. The specific purpose in this section is to clarify the priorities for land use zoning in order to strengthen participatory flood and water management approaches. To this end, both land use planning and water management would have to be integrated as one cohesive plan through thorough coordination among various land and water users and related agencies. With consistency in planning, authorities will be efficiently coordinated by sharing responsibilities among themselves. To gain further insights, the main procedures to address in this part are discussed in the following.

Land use zoning

Zoning is a land use control that proposes the type and density of development within a specific area. It includes a process of determining the most desirable way for how land should be used under the condition that the plan needs to be based on legally defensible and enforceable land use plans. The purpose of zoning in Chi River Basin is to offer a balance of different land uses such as recreation, environmental conservation, agricultural, residential, commercial, industrial, etc. and also to reduce potential conflicts between incompatible land uses, in order to ensure the public health, safety, and general welfare. According to the description mentioned above, land use zoning can also be used to mitigate flood disasters and reduce risks by directing development sites not to be in hazard-prone areas, i.e. by using better site designs, promoting developments situated in specific areas, limiting/prohibiting development in some areas, and providing protection for priority land conservation areas.

In considering land use zoning, it is necessary to converse the land to a conforming use and divide the areas into zones with clear development criteria of each zone to minimize the hazard vulnerability. In the end, the following zoning categories would have to be taken into consideration:

- urban residential zone, the proposed approach to urban residential zoning responds to the need to increase flexibility in building form, reduce the number and complexity of regulations that obstruct residential intensification efforts, and reflect a greater design approach to community planning;
- agricultural zone, the purpose of this zoning is to preserve the existing agricultural uses. Moreover, supplementary uses that are not directly associated with agriculture are also open for consideration;
- industrial and warehousing zones, this zoning provides for low-density employment uses such as industrial and warehousing uses. Other uses supplementary or similar to industry and warehousing business are open for consideration based on the benefits of individual planning application and may be acceptable in this zone;

- open space, park, and amenity zones, the specific aim of this zoning category is to protect and provide for recreation, open space, park, and amenity provision to the community. This zone relates to both public and private open space, while its zoning objectives are dispersed throughout a town. The officials as a rule should not permit any development project that would result in a loss of established open space and park except where specifically provided in the development plan.

In addition to this, to find out the use of zoning as the way to achieve effective land use planning, the further procedures involved, which includes information management through GIS, spatial-temporal modelling on present land use, alternative scenarios, and assessment of consequences, are necessary.

Water management systems zoning

Water is essential for survival, too little or too much water can be life threatening. Proper management is necessary for water to be delivered where and when it is needed. It can also help to mitigate the problems of overabundance like floods. In the Chi River Basin, seasonal variation in rainfall with most of the rain falling during 4 to 5 monsoon months creates problems of sustaining people's livelihood, hence well planned water management is strongly needed. The water management system needs to be established to support the water control management mission in order to acquire, transform, verify, store, display, analyse, and disseminate data and information efficiently and timely. Water management systems zoning consists of the following:

- the incoming data, which includes hydrological information, i.e. river stage, reservoir elevation, meteorological information such as rainfall and static information namely sub-basin area, reservoir storage capacity, channel geometry, spillway elevation, etc, will be collected and summarized in several formats, e.g. graphs, tables, spreadsheets, charts, river profiles, maps, etc. The results will then be used in water management system analysis in order to process, store, and made available for viewing and river basin modelling;
- the river basin processes will be simulated; river basin modelling which represents the hydrological and hydraulic aspects of the river basin will be used to forecast runoff, reservoir response and operations, river stage, inundated area, and downstream impacts;
- GIS tools will be used to carry out analyses and enhance graphical displays. Results of the simulations will be further used to map inundated areas. As a result, inundated area will be combined with other GIS data to identify critical locations, which are possible to be affected by the event such as residential areas, roads, and other infrastructure, etc.;
- the simulation results will be used to define appropriate reservoir operation rules and policies together with the knowledge of how those selected operations will affect both in-lake and downstream areas;
- the parameter settings, such as reservoir operation and different future rainfall amounts, will be modified to produce a range of possible scenarios. Consequently, the results and impacts will be compared and evaluated;
- it will describe how information can be disseminated to different interested parties including direct and indirect users, cooperating agencies, interested organizations, and individuals;
- as a result, a water management systems zoning will be obtained by applying a spatial integrated technique as shown in Figure 5.15, which will be used as a decision supporting tool to flood risk management strategies.

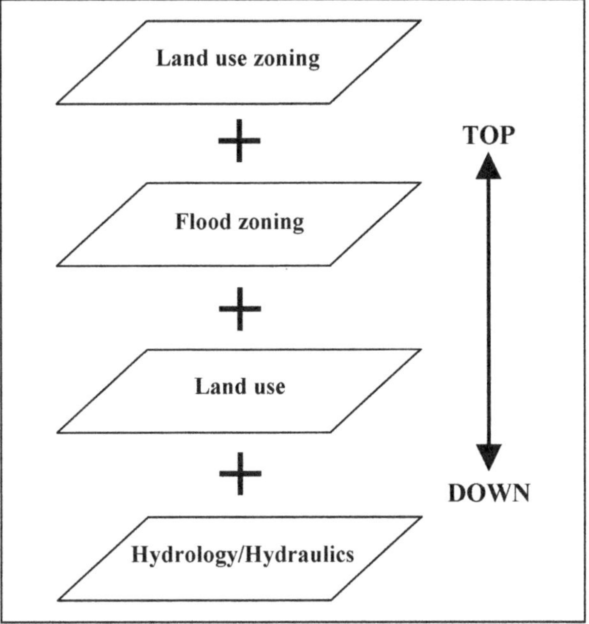

Figure 5.15 Spatial integrated technique for flood management strategy

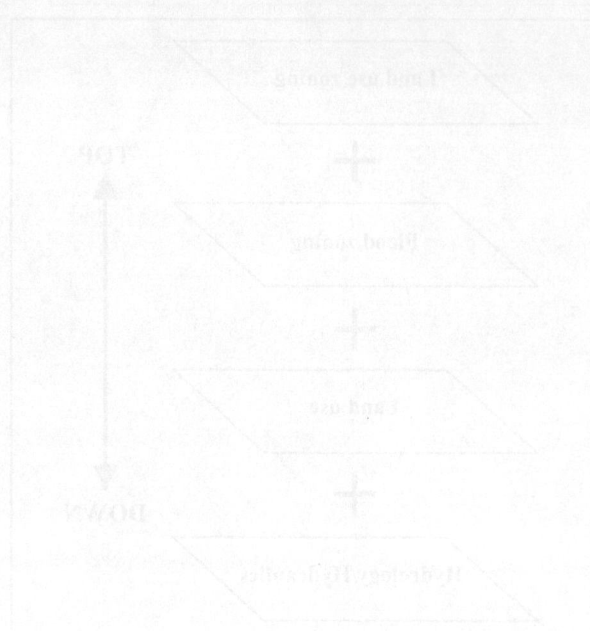

6 Application of mathematical models for flood investigations

6.1 Overview

When assessing flood vulnerability, unanswered questions regarding a clear view of what happened before, during, and after the flooding may still remain. These questions may seem simple to answer, however, there is no simple answer nor simple solution for it. To respond to these frequently asked questions in strict confidence, more detailed investigation and examination will normally be required.

Technical investigation of flood characteristics could help to reduce future loss and damage, as well as to manage and control future development within the river basin. With conjunctive flood investigation, it also offers an effective way to tackle the questions on the level of mitigation and consequences of flooding, as well as to raise awareness and to seek for careful structural and non-structural measures. On the basis of investigation around a specific point of interest, mathematical models can be used in order to gain an improved understanding of the hydrological and hydraulic interactions, in particular flood regime changes in the river system under man's intervention. Likewise, the investigation of hypothetical scenarios is also a crucial and active research area to span a dynamic perspective through a range of future possibilities with development assumptions. Under this circumstance, one should question what types of changes might be needed to ensure that a catastrophe will not recur in the future. Such processes invariably cover different aspects of applications related to engineering solutions, it is assumed that any flood management technique would be located and alleviated any adverse flood impact, as well as not to inadvertently create any problem elsewhere in the system. The creation of different flood scenarios is undertaken to investigate the effectiveness of physical flood mitigation measures to reduce the amount of damage. Moreover, any proposed solution also needs to take the following two referral questions into account: how does the flood mitigation system perform? And, what will be the successes and shortcomings? Most importantly, it is also a prudent initial step to reduce future investment on flood management by means of an integrated approach. It is not necessary to fix options as it might to some extent minimize the ability to adapt at a later date. In all this, however, no matter how extensive and thorough the investigations and flood management through engineering solutions are, some potential risk will always remain.

6.2 Mathematical model applications

Before envisaging a successful solution to untangle the issue related to water resources management problems, every single answer would need to be considered whether it is an appropriate solution for the Chi River Basin. Any advice on what to do and what impacts may result will inevitably involve the modelling of the virtual river system with properties similar to the actual one.

The subsequent sections examine the application of detailed models used to tackle flood threat and its interaction with other processes in this particular area. More will also be said about the assumptions that allow for what-if analyses of potential development scenarios whereby its results are used as technical assistance in designing a flood action plan.

6.2.1 SWAT model application

Prior to putting any effort in water resources management, there is one hydrological question that is worth asking, i.e. what will happen within a river basin in the future? In light of this, everything that is mentioned in this section is an attempt to better understand this curious statement and its influence. Such efforts would likely be directed towards the possible consequences of future hydrologic conditions as of particular concern to the rainfall-runoff phenomena. In response to this concern, the application of a process-based hydrological model such as SWAT is indispensable. The following describes the main components, which are deployed to develop interactive applications. The steps involved are given in Figure 6.1.

Input data used

On the basis of the physical rainfall-runoff transformation process, the input data set is considered to be of primary importance to a successful simulation application. The input data requirements of the SWAT model are composed of the following:

- *digital elevation model.* Spatial information for the model is provided by a 300 * 300 m^2 Digital Elevation Model (DEM). The DEM was used in determining the flow direction for each raster cell, which closely followed the flow pattern suggested by the DEM. In addition, DEM was also used to derive the drainage networks and eventually delineates the sub-basins in order to provide an adequate representation of the topography. Initially, the DEM was derived based on the product of the Shuttle Radar Topographic Mission (SRTM; 90 * 90 m^2) and further topographical data available from Royal Thai Survey Department (scale 1:50,000);
- *land use data.* The land use patterns were derived from the Thailand Land Development Department data sets (2001). These data sets were received in the form of a digital map (scale 1:50,000), with polygons representing different land use types. Processing of the data was done by using ArcView GIS, and categorizing known land uses into larger groups, which were matched to the appropriate land uses included in the data base of SWAT. Eventually, a total of 18 different land uses were chosen for this study;
- *soil data.* The soil data were also gathered from the Thailand Land Development Department (2001) at 1:50,000 scale and clipped to fit the domain of the Chi River Basin. The upper soil layer is used in SWAT for calculating infiltration, ponding and runoff generation. The 37 different types of soil that exist in the river basin were introduced into SWAT in the form of a digital map. A series of attributes were assigned to each type of soil, e.g., depth, saturated hydraulic conductivity and soil texture. The soil pattern in the river basin is one of the vital input data sets to the model. It is combined with the land use data to determine areas with similar hydrological response (HRU), without reference to its actual spatial position within each sub-basin (semi-distributed set-up);
- *hydro-meteorological data.* The meteorological data were derived from 8 meteorological stations and 76 rainfall stations in and around the Chi River Basin (Figure 6.2). These data also include daily temperature, average mean monthly values of wind speed, solar radiation and relative humidity, corresponding to the period 1980-2005. While time series of streamflow data were derived from hydrological stations. The daily streamflow data at the Ban Tha Khrai gauged site

(E18) for the period June 1 - October 31, 2001 were used for the calibration of the model.

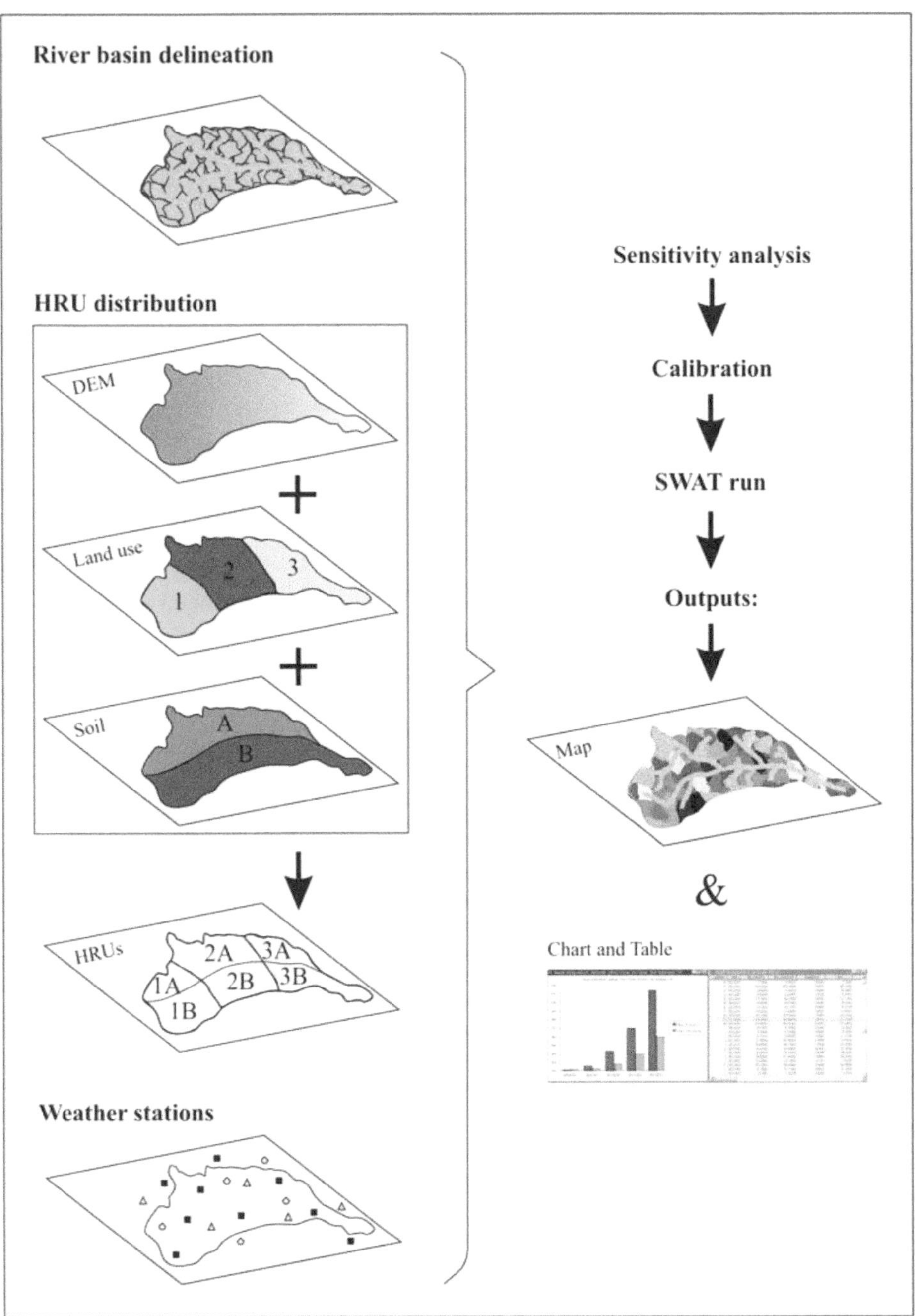

Figure 6.1 Steps involved in river basin simulation with the SWAT model

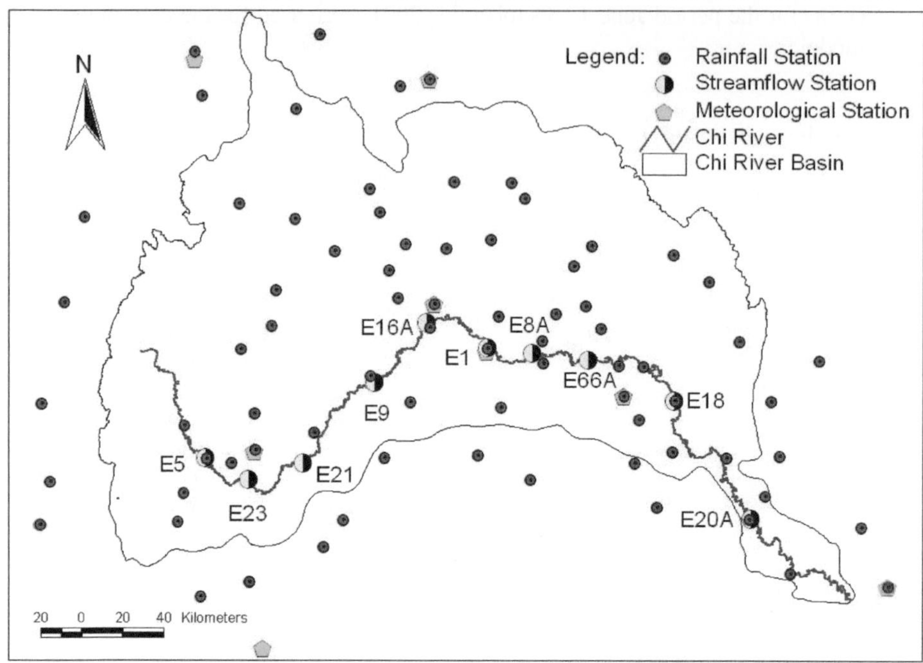

Figure 6.2 Hydrological and meteorological instrumentation sites used for SWAT model application

Model implementation and sensitivity analysis

The SWAT model was tested for identifying and quantifying a provision for channel routing of runoff from tributary sub-basins. The daily SWAT simulation covered the period 1 January 2000 - 31 December 2001. This period was chosen because of the availability of input data. Moreover, the chosen simulation period was also based on the historical data, as the floods of 2000 and 2001 were among the most devastating floods in the history of the Chi River Basin. Regarding the frequency analysis of maximum discharges at the Maha Chana Chai station (E20A) (the downstream part of the Chi River Basin), the 2001 flood corresponded to the 4% annual probability of exceedance (Pawattana et al., 2007). From the reasons given above, hydrological data in 2001 were used for model calibration.

Although the start date for all model simulations is 1 January 2000, however, model outputs are only displayed for the calibration period 1 June - 31 October 2001. In this regard, a one and half year initialization or start-up period was used, so the impacts of uncertain initial conditions in the model are minimized. In addition, as no daily time series data were available at the outlet of the Chi River Basin, calibration was undertaken at the Ban Tha Khrai gauged site (E18) within the river basin (Figure 6.2). Before calibration, a sensitivity analysis was performed in order to determine the parameters to which the model results are most sensitive. Consequently, these parameters were then used in the calibration process in order to obtain good discharge simulations at station E18. Thereafter, the calibrated model parameters were used to simulate streamflow using different rainfall scenarios for the predictions of the tributary inflows with a specified probability of occurrence at all the selected points along the Chi River. A ranking of the most sensitive parameters is given in Table 6.1.

Table 6.1 Ranking of the six most sensitive parameters and their variation range for autocalibration (in alphabetical order)

Parameter	Description	Unit	Range	Process
ALPHA_BF	Baseflow alpha factor	day	0 - 1	Groundwater
CH_K2	Effective hydraulic conductivity in main channel alluvium	mm/hr	-0.01 to 150	Channel flow
CN2*	SCS runoff curve number for moisture condition II	-	-25% to 25%	Runoff generation
ESCO	Soil evaporation compensation factor	-	0 - 1	Evaporation
SOL_AWC*	Available water capacity of the soil layer	mm H_2O/mm soil	-10% to 10%	Soil
SURLAG	Surface runoff lag coefficient	-	0 - 10	Runoff generation

Note: * These distributed parameters were changed by a certain percentage, i.e. relative change (±10% and ±25%), whereby the HRUs initially having the highest (lowest) parameter values will remain having the highest (lowest) values, in order to maintains their spatial relationship

The SWAT model generally has a large number of parameters to capture the various physical characteristics of a hydrologic system. Manual calibration of such model is a very tedious and daunting task, and its success depends on subjective assessment with knowledge of the basic approaches and interactions in the model. In order to alleviate these shortcomings, an automatic calibration procedure called PARASOL (Parameter Solutions Method) (van Griensven and Meixner, 2007) was applied to each sub-basin separately after some first manual calibration and good estimation of the possible parameter ranges. The PARASOL method operates by a parameter search method for model parameter optimization, the so-called Shuffle Complex Evolution algorithm (Duan et al., 1992). Subsequently, a statistical method was then performed during the optimization to provide parameter uncertainty bounds and the corresponding uncertainty bounds on the model outputs (van Griensven et al., 2006).

Model calibration and validation

While there is truth to the widely used SWAT model that cannot be denied in which there are many parameters to be testable and some of them are generally not available. Thus, it seems fairly clear that the model would first have to be calibrated to measured data in order to determine the best or at least a reasonable parameter set.

As it proceeds, the SWAT model is subject to a two-stage calibration and validation procedure. Whereas model calibration was based on the wet period from June 1 through October 31, 2001 and validation was based on another wet year (2002). Indeed, one may argue that the calibration period is too short. However, the chosen calibration period does contain at least one extreme event towards the end of the 2001 flood, and revealed promising results. Regarding the start-up period of the model, it is rather long in comparison to the calibration period (Figure 6.3), as it needs to be long enough to initialize a stationary condition. Following this start-up period, a prolonged calibration period for a few more months by deducting from the start-up period was omitted, since it did not change the parameterization of the SWAT model.

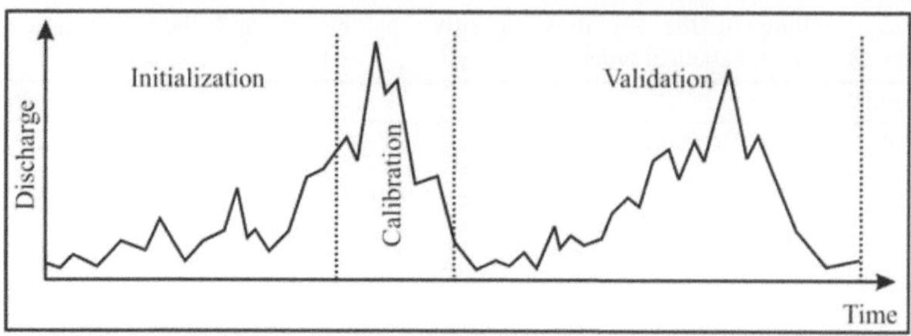

Figure 6.3 Selection of calibration and validation intervals for SWAT model

With respect to the calibration results at the streamflow station E18 as shown in Figure 6.4, the model performance is satisfactory. This assertion is confirmed by a suite of evidence indicators. In this perspective, several statistical measures were used to evaluate the simulation accuracy (Table 6.2), such as the Nash-Sutcliffe coefficient (E_{NS}) to indicate how well the observed and simulated values fit each other (Nash and Sutcliffe, 1970), Root Mean Square Error (RMSE) to quantitatively measure how closely the simulated values track the actual data, Goodness of fit (r^2) as an indicator of strength of relationship between the observed and simulated values, and the Mean Absolute Error (MAE) to determine the average over the verification sample of the absolute values of the differences between simulated and corresponding observed values. If the E_{NS} and r^2 values are less than or very close to zero, the model accuracy is considered 'unacceptable or poor'. If the values are one, then the model prediction is 'perfect' (Santhi et al., 2001). In general, SWAT model performance is considered satisfactory or acceptable if E_{NS} and r^2 values greater than 0.5 (Moriasi et al., 2007). In such instances, it is obvious that E_{NS} and r^2 values are high, which indicates that the SWAT model is able to produce consistent results in estimating tributary inflows at all the selected points on the Chi River, under given climatic conditions. After that these tributary inflows can become a key input into the 1D/2D SOBEK hydraulic model.

After obtaining a satisfactory calibration, the hydrological components of SWAT were further validated with one additional year (2002) of weather and streamflow data, whereas a set of input parameters used for the calibration remained unchanged. This period was chosen because it includes the extreme flood event in September 2002, as well as a dry period with severe drought.

To be prudent, graphical examination and descriptive statistical measures were used herein to verify the performance of the model. The graphical result shown in Figure 6.4 represents the tendency of slight under/overestimation of the simulated hydrograph. This bias estimation is probably because of local rainfall that was not well represented by the input rainfall data used in the SWAT simulations. It is also associated with parameter calibration of the runoff simulation process, SCS curve number (CN2), which strongly controlled the rate of surface runoff generation. In spite of its overestimation of peak flow, however, the overall graph shows a good agreement for the independent data set with measured and simulated daily flows.

Besides the graphical examination, the same standard statistical techniques were also used in the calibration process. In general, model validation is judged satisfactory, this evidence is supported by good performance from statistical comparisons of validation results with observed data (Table 6.2). Although the E_{NS} and r^2 values are somewhat less than those obtained from calibration, they are still higher than the recommended minimum values quoted in the literature.

Upon successful completion of the processes, it is fair to claim that once properly calibrated, the SWAT model can adequately simulate and capture streamflow dynamics in the Chi River Basin.

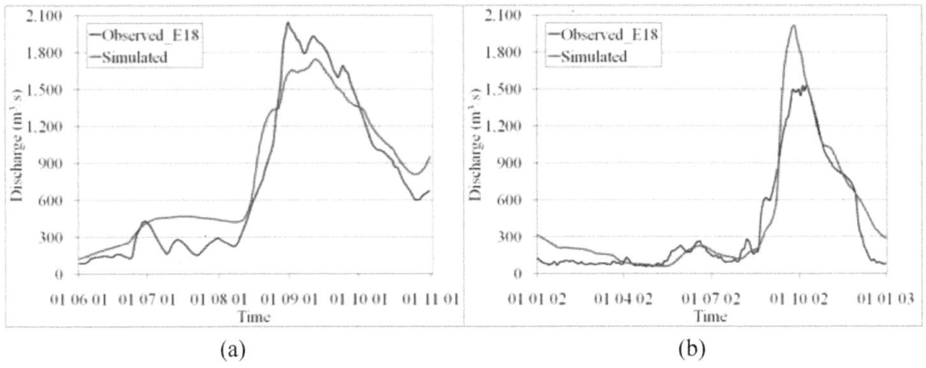

Figure 6.4 Comparison of SWAT simulated and observed daily discharges during: a) calibration period; and b) validation period

Table 6.2 Statistical indices of SWAT model performance calculated during calibration and validation phases at streamflow station E18

Process	Period	Index			
		E_{NS}	RMSE	r^2	MAE
Calibration	June - October 2001	0.91	185	0.95	159
Validation	2002	0.86	166	0.91	111

6.2.2 1D/2D SOBEK model application

For the application of the 1D/2D SOBEK, the model needs to be set up in order to identify the propagation of floods through floodplain areas with a time step of 30 minutes (rather than daily) to deal with flood simulations. The hydraulic system of the Chi River was schematized by using the following inputs:

- *1DFLOW module*. GIS layers were imported to represent the stream network and lateral inflow points. These layers were derived from 1:50,000 digitized survey maps.
- *overland flow (2D)*:
    - the topography was defined using a 300 * 300 m^2 Digital Elevation Model (DEM), based on detailed topographic maps, 30 * 30 m^2 DEM and 90 * 90 m^2 SRTM data. It has to be noted that the effects of the grid resolution and vertical precision for the success of hydrologic and hydraulic modelling have not been evaluated as they were not the main purpose of this study. Somehow, it seems that these are not the real factors controlling the quality of terrain parameters but need to be inferred from the scale of research and quality of the given topographic data. The DEM covered approximately 49,500 km^2 and served as input to the flood simulations;
    - boundary conditions in the form of discharge at the upstream and water level on the downstream side;
    - tributary inflows at all the selected points on the Chi River as a result of SWAT simulations and used as the input (coupled) with the 1D/2D SOBEK model for flow routing;

- river cross sections at different locations from the streamflow station Ban Non Puai (E5) through the Chi-Mun confluence;
- time series of 1980 - 2005 of water levels and discharges at various streamflow stations;
- estimated roughness coefficients for the main channel, which were obtained by calibrating the roughness in the river channel using 1D modelling of rivers/channels (1D SOBEK). The 0.100 s/m$^{1/3}$ floodplain roughness was derived from the available literature. Meanwhile, it was assumed constant throughout the floodplain surface due to the unavailability of detailed spatial roughness information. Of course, this parameter will influence the hydrograph timing and peak, and is subject to calibration. Thus, in future, more details of spatially distributed Manning's roughness coefficient for the floodplain need to be investigated.

Calibration and validation of 1D SOBEK river flow model

To start the process of model calibration, it was essential to define the model parameters of interest. In response to this concern, the Manning's roughness coefficient, which is the most important parameter that reflects the resistance against flow along a channel, was the only parameter used in this (subjective and) time-consuming process. In making 1D SOBEK simulation looks real, the model was calibrated using water levels of various streamflow stations for the extreme flood event of 2001. A ten-month period was used for the calibration, from January to October 2001. However, the first three-months were not taken into account since it was considered as a warm-up period to initialize the model variables, as well as to ensure that the model reaches a realistic condition (Figure 6.5). Ideally, through adjustment of the parameter, simulated water levels were expected to be fit as accurately as possible onto the observed ones. The Manning's n value was derived from Chow (1959), and was then modified subsequently. Changes were made in channel roughness, while for the adjacent riparian zone (floodplain) a constant value was assigned as described earlier. In order to bring the channel roughness, which may vary between locations into harmony with the real situation, the model calibration focused on calibrating three sets of Manning's n value throughout the routing reach, but within reasonable ranges (Table 6.3 and Figure 6.6). And the comparison of calibration results was done with more than one calibration fixed point, i.e. Ban Tha Khrai (E18) and Maha Chana Chai (E20A) gauged sites.

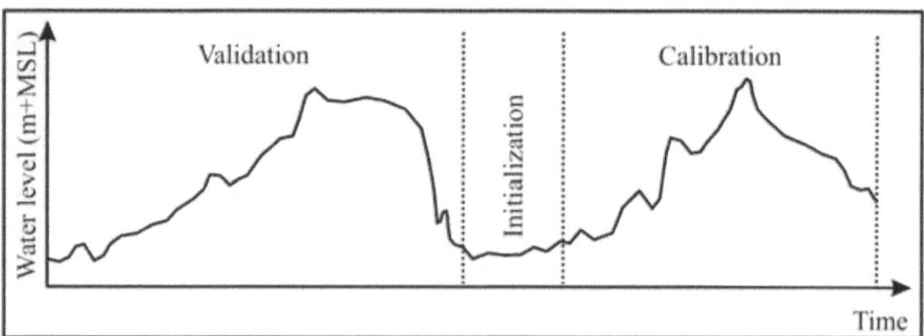

Figure 6.5 Selection of calibration and validation intervals for 1D SOBEK model

Upon successful completion of calibration, the calibrated channel roughness coefficients, which bridge the gap between model and data, are given in Table 6.3. The ultimate results from the calibration runs based on different roughness values were eventually achieved. Figure 6.7 illustrates the comparison of observed and simulated water levels at the two gauging stations E18 and E20A. Based on visual judgement from those two illustrated figures, there is a reasonably close agreement of reference data and model results, particularly during peak flow conditions.

Table 6.3 Manning's roughness coefficients after the calibration

Reach	Location	Manning's n in $s/m^{1/3}$
1, 3, 4, 5	Ban Non Puai gauged site (E5) - Chi-Nam Phong confluence	0.045
6, 7, 8	Chi-Nam Phong confluence - Chi-Nam Yang confluence	0.032
9	Chi-Nam Yang confluence - Chi-Mun confluence	0.028

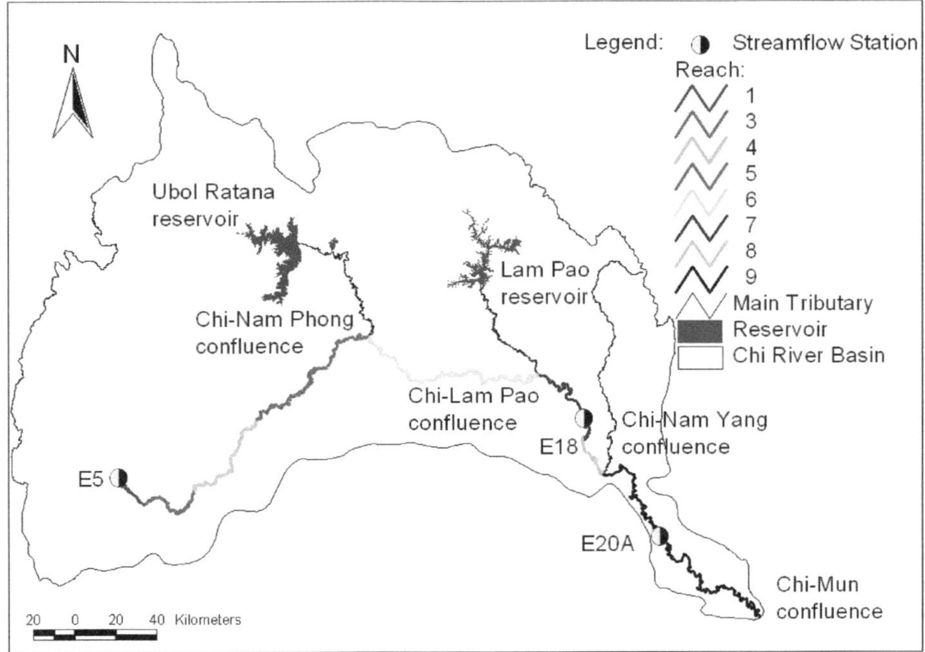

Figure 6.6 Reach locations for identification of calibrated Manning's roughness coefficients

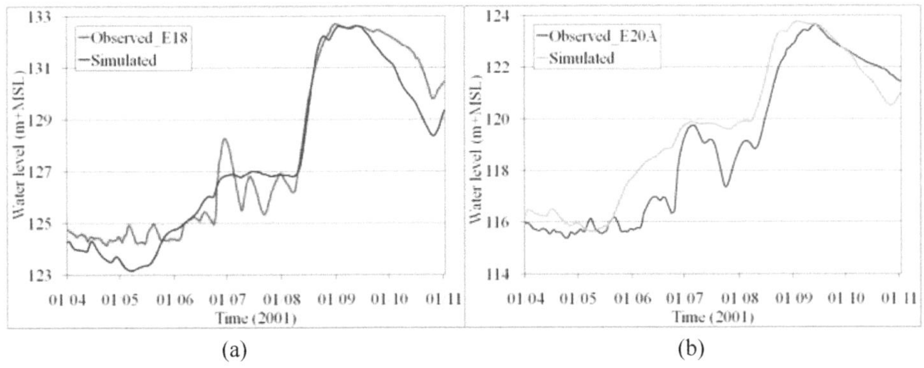

(a) (b)

Figure 6.7 Comparison of 1D SOBEK simulated and observed daily water levels at: a) Ban Tha Khrai (E18); and b) Maha Chana Chai (E20A) during the calibration period

Other than the qualitative closeness relationship through visual inspection, the goodness of fit would also have to be examined quantitatively. In respect to this, performance statistic, i.e. goodness of fit (r^2), is used to assess the performance of calibration results of 1D SOBEK for different water level gauging sites as presented in Table 6.4. With the iterative calibration process, simulation with the adjusted Manning's roughness coefficients varying from 0.028 s/m$^{1/3}$ to 0.045 s/m$^{1/3}$ was found to be able to reach the highest r^2 values at both calibration points. As of this finding, it is a good sign of confidence to claim that 1D SOBEK can perform reasonably well during the calibration process.

Table 6.4 Statistical performance index for different water level gauging stations during the calibration and validation period

Calibration point	Performance index (r^2)	
	Calibration	Validation
Ban Tha Khrai (E18)	0.94	0.79
Maha Chana Chai (E20A)	0.91	0.89

To come up with the drawing of realistic conclusions from which the 1D SOBEK model reflected the real conditions, the validation was conducted using an event (as of 2000) that differs from those used for calibration. The results described and illustrated here are visually apparent in Figure 6.8. The presence of this finding clearly points to their reasonable trends, which are generally in similar agreement with the measured water level values at both validation sites, despite some local under/overestimation. The effect of Nam Phong tributary inflow (water released from Ubol Ratana reservoir) on conditions in the Chi River is the main reason that the validation results at the gauging station E18 go beyond the observed data. To ensure confidence in the results, the performance of daily validation estimates was evaluated using statistical indicators, as summarized in Table 6.4. From this table, it is observed that the goodness of fit (r^2) in both validation data sets is quite satisfying. This confirms the graphical results and emphasizes that the calibrated model performed reasonably well during validation.

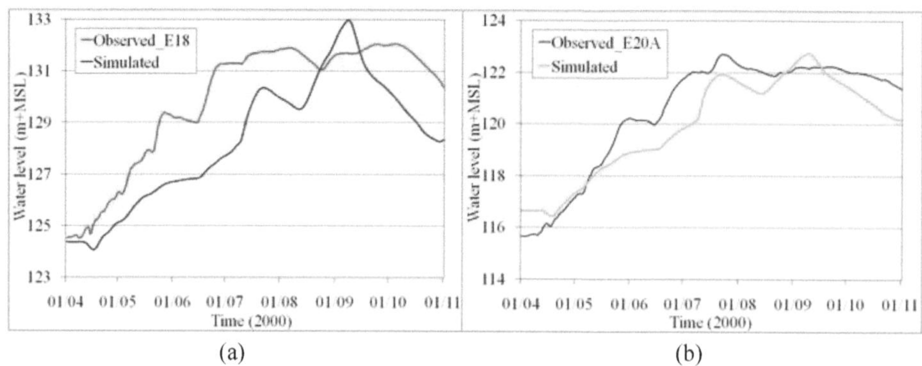

(a) (b)

Figure 6.8 Comparison of 1D SOBEK simulated and observed daily water levels at: a) Ban Tha Khrai (E18); and b) Maha Chana Chai (E20A) during the validation period

After completion, the promising results suggest that 1D SOBEK can be used to serve as a starting point for the flood propagation modelling along the Chi River and its tributaries.

6.2.3 Comparison of spatial patterns of flood inundation

To ensure a successful calibration as described above, the calibrated 1D SOBEK model was dynamically coupled with the 2D hydraulic flood propagation component (2D SOBEK) using a constant value of Manning's n equal to 0.100 s/m$^{1/3}$ for floodplains as mentioned earlier. The simulated flood inundation result obtained from 1D/2D SOBEK was then validated using the actual inundation extent derived from RADARSAT satellite imagery dated 13th September 2001 that served as a benchmark to assure accurate and reliable results. In this comparison, the primary focus was the determination of flood inundation areas as a result of water spilling from the Chi River and flowing along the floodplain, and a meaningful comparison was carried out based on the spatial extent of the same date. The comparison was classified into three classes, i.e. the close match (identical) in both RADARSAT image and simulation result (red), under-prediction by model: areas that appear wet on the observed map but show dry on the simulated map (black), and over-prediction by the model: areas that display dry on observed map but present wet on simulated map (green). At first sight, the simulated flood inundation extent is consistent with the expected pattern and is in reasonable agreement with the corresponding actual inundated area (Figure 6.9). It is noticed that the vast majority of destructive floods coincidently occur in the downstream part of the Chi River Basin, which is in line with the absence of a downstream portion of the dike (see Section 3.4.3 for details), as well as flat and low-lying topography.

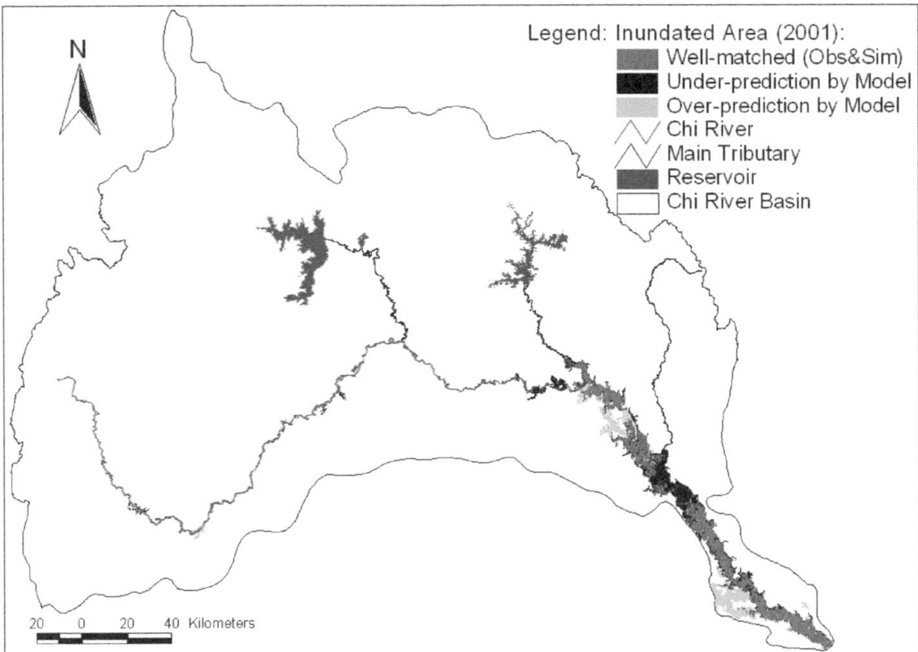

Figure 6.9 Comparison of spatial inundation patterns resulting from satellite-based estimates and model simulation

From visual evidence, there is a slight difference in terms of total flood extent, i.e. the model simulated 1,100 km^2 as flooded areas, while the observed extent encompassed approximately 1,330 km^2. This difference may be attributed to the fact that the calibration of floodplain roughness in this study was not considered in which the consideration of this task is expected to improve the results. Another reason perhaps because of the presence of tributaries in the downstream areas of the Chi River, which has not been considered in the 1D/2D SOBEK model but causes waterlogging and flood inundation.

Overall, the performance of the 1D/2D SOBEK hydraulic model was found to be satisfactory as it was successfully applied and capable of simulating and providing accurate flood inundation extent for the Chi River Basin. Due to this, the simulation of this calibrated and validated model with different annual probability of exceedance of floods may generate practical approximate predictions in novel situations, which cover a range of model scenarios for the purposes of flood mitigation as well as other management works for the Chi River Basin.

6.2.4 Mean areal rainfall

The thought on the fact that rainfall is a significant aspect of hydrologic response, particularly, it could potentially exacerbate future flooding problems. In response to this concern, the rainfall measured at a point was converted to an areal estimate at a daily interval for the computation of a design storm with a specified annual probability of exceedance by using a rainfall data set from the Thai Meteorological Department (TMD). There are 105 stations with daily rainfall records in and around the Chi River Basin, ranging from 1951 - 2006, however, some gauges only have short records. As the distribution of rainfall varies over space and time, it is required to analyse the data covering long periods and recorded at various locations to obtain reliable information. To ensure statistical stability of the results, the selection has been made to choose gauges with at least 20 years of records up to 2005. Then fifty-eight stations were chosen (Figure 6.10).

In an attempt to obtain the most likely spatial distribution of rainfall, Thiessen interpolation technique was used by which it divides the Chi River Basin into polygons containing the rain gauge in the middle of each polygon (Figure 6.10). The rainfall at any location within the same polygon is assumed to be the same as the concerned rainfall station. In the construction of Thiessen polygons, Geographic Information System (GIS) was used. Whist the mean areal rainfall across the Chi River Basin can be calculated by multiplying rainfall associated with that polygon by the area of the polygon and dividing by the total area (Equation 6.1).

$$\overline{P} = \frac{\sum_{i=1}^{n}(A_i P_i)}{\sum_{i=1}^{n} A_i} \tag{6.1}$$

where:

$\overline{P}$ = mean areal rainfall over a river basin (mm/day)
A_i = area of polygon belongs to rain gauge i (km^2)
P_i = amount of rainfall at rain gauge i (mm/day)
n = total number of rain gauges (-)

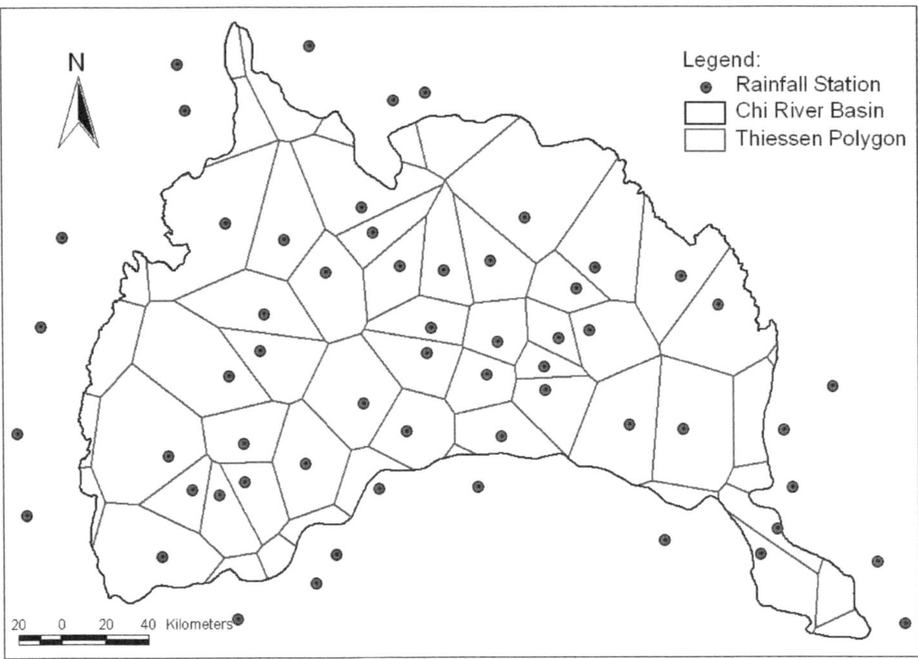

Figure 6.10 Location of rainfall stations and coverage area of each rainfall station determined by Thiessen Polygon

6.2.5 Statistical estimation of rainfall

Exploration of the statistical properties of hydrologic variables, especially rainfall, is considered important in the identification of future hydrologic phenomena, since the characteristics of the climatic processes are assumed to remain unchanged. As a matter of fact, rainfall events occur randomly, however, within a long period of analysis of distribution, it indicates that the frequency of occurrence of events tends to follow statistical patterns.

In this context, the statistical estimation of rainfall is addressed, in which it is essential and fundamental when viewed in the context of rainfall-runoff processes. Under this basis, the 2001 flood in the Chi River Basin, has been considered to determine the base condition rainfall for the flood hydrograph. Since the tropical storm Usagi in 2001 lasted 3 days, the rainfall analysis is performed at 3-day time interval.

In doing this statistical matter, analysis of rainfall data requires handling of large volume of data and repeated computation of a number of statistical parameters for distribution fitting and estimation of expected rainfall at different annual probability of exceedance. The quantitative estimation of rainfall for a given frequency is commonly done by utilizing different probability distributions. Different distribution functions are mostly approached with the Gumbel Type I distribution (also known as the Extreme Value Type I distribution), log-normal, Pearson type III (Poisson gamma distribution), log Pearson type III, and Gamma distribution. At this point, Gumbel's extreme value distribution has been applied to the analysis, because it fitted the data the best.

To find the maximum rainfall of different annual probability of exceedance for a duration of 3 days, the Gumbel's extreme value distribution has been applied to the annual extreme rainfall series of 3-day duration for 26 years (1980 - 2005). In this respect, the Gumbel's Extreme Value Statistical Distribution equation is given by (De Laat, 1992):

$$X = \overline{X}_{ext} + \frac{s_{ext}}{\sigma_N}\left(y - \overline{y}_N\right) \tag{6.2}$$

$$y = -ln\left(-ln\left(1 - 1/T\right)\right) \tag{6.3}$$

where:

X	= unlimited exponentially-distributed variable
$\overline{X}_{ext}$	= mean of the independent extreme values
s_{ext}	= standard deviation of the sample
σ_N	= standard deviation of the reduced variate
y	= reduced variate
$\overline{y}_N$	= mean of the reduced variate
T	= return period

The calculated extreme rainfall has been subsequently placed at the highest three-day mean areal rainfall in 2001 estimated by Thiessen polygon method, whereby the amount for each day was varied based on the distribution of those three-day peaks and no other changes were made to the remaining part of the time series (Table 6.5). (Note: it was assumed that rainfall intensity is uniformly distributed over a time period across the entire Chi River Basin, even though the study area is large and rainfall intensity will vary spatially and temporally during a storm). Afterwards, the design rainfall was then put into a distributed rainfall-runoff model to generate the flood hydrograph for extreme conditions that may result in flooding.

Corresponding to the estimation of rainfall for such a high probability of exceedance, the process involves extrapolation far outside the range of observed data. It is generally recommended that data can accurately extrapolate up to twice the length of record, while most of the time extrapolation goes beyond twice due to inadequate data. However, extrapolation can still be made, but it is subject to a wide margin of error.

Table 6.5 Daily areal rainfall estimation for the specified annual probability of exceedance event

	Areal rainfall		
Date	for 2001 in mm/day	distribution in 2001 in %	for 1% annual probability of exceedance in mm/day
9/8/2001	29.4	34.2	60
10/8/2001	27.1	31.6	55
11/8/2001	29.4	34.2	60
Total	85.8	100.0	175

6.2.6 Design flood modelling

In recognition of the reality, flood events are generally beyond human experience. To continue along the line of insight investigation, the importance of dynamics of flood propagation is now well recognized and more emphasized. In other words, much interest is being paid to the processes for design flood estimation.

A design (hypothetical) flood and its consequences is the key aspect in hydrologic design of engineering practices. It represents an extreme scenario likely to occur at a specific likelihood of occurrence. In this regard, the design flood considers the degree of hazard on a case-by-case basis. The selection of an appropriate design flood is of great importance, while underestimation might cause devastating effects on life and property. In relation to the concept of design flood in the Chi River Basin, the exceedance frequency (safety standard) has been fixed at 1% annual probability of exceedance at a given location where many people might be potentially affected.

In relying on the criterion 'an annual probability of exceedance of the estimated flood is equal to an annual probability of exceedance of the input rainfall', the design flood event was determined using SWAT and 1D/2D SOBEK to determine the corresponding design flood.

6.3 Ways of managing floods

Whilst floods are often considered amongst the most manageable of the natural threats (Keys, 2004), as it can be anticipated regarding where and when floods will occur including what their consequences will be. However, intensive investigations of flooding opportunities in order to alleviate their impacts are still ongoing to maintain them within manageable levels. The manageability of flooding is achievable, although not at all easy, by deploying state of the art engineering solutions that are increasingly being enquired and indeed exposed (Figure 6.11). In consequence, a few papers currently focus more on the management of flooding. As stressful as this situation is, flood management is obviously crucial and therefore deserves priority attention to fight against floods.

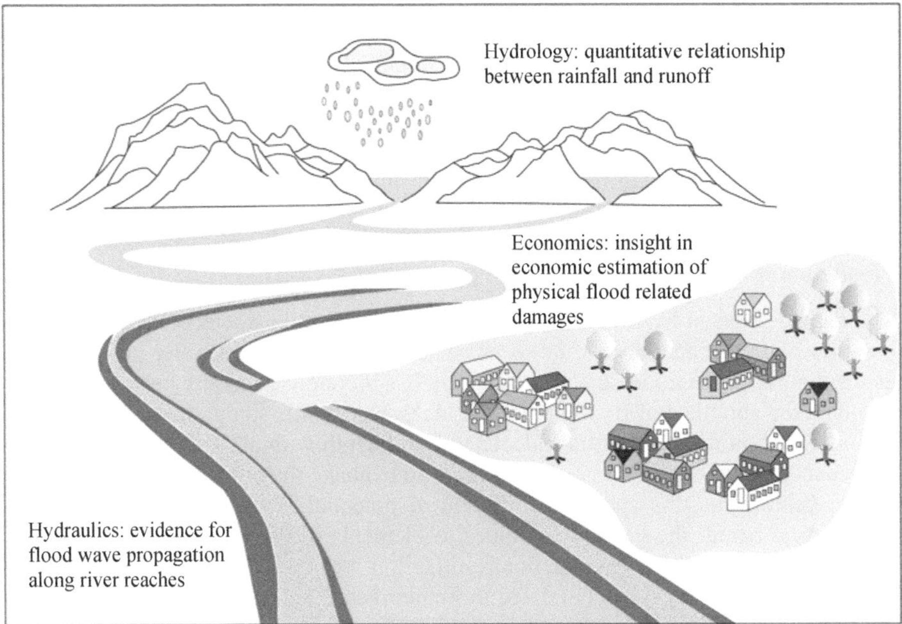

Figure 6.11 Key contributing factors involved in flood risk management system

6.3.1 Scenario development

Taking into consideration the scenarios for future floods, the better understanding on the relationships among human activities and flood occurrence will allow water authorities to make better comprehensible decisions on flood control and management.

Needless to say, scenarios are definitely useful and necessary for investigating potential flood management strategies to cope with different future hypothetical situations. Different alternatives produced by different simulations are in fact a consequence of applied scenario changes, and it can be applied to all scenarios. In this study, flood mitigation scenarios were investigated with the model for assessing the changes in flood risks and the likely impacts, in order to come up with a strategic plan to be included in a comprehensive programme for integrated flood risk management in the Chi River Basin. However, it should be noted that it is not necessary for this study to cover development scenarios and cases outside the scope of the study. Moreover, it was also not considered to be economically viable to provide flood mitigation measures that would alleviate the flood prone areas affected by the previous severe floods for a future event with similar magnitude. Therefore, the potential hydraulic impact of a flood mitigation effort was investigated for the estimated 1% annual probability of exceedance flood, which is the concept design standard in Thailand for areas with developments.

6.3.2 General analysis of alternative flood management practices

It is true that some of the possible flood mitigation actions shown at the top list are expected to deliver the desired outcomes. However, what is the indication that alternatives to what is proposed are likely to provide adequate justification for choosing those options. To determine which flood management practices would be successfully implemented, a sound basis for financial justification is necessary.

Financial analysis consists of estimating and comparing the costs and benefits. To be financially attractive, the flood damage reduction (benefit) provided by a particular mitigation option needs to exceed the implementation cost, which includes its construction, operation, and maintenance expenses. Detailed comparisons have been made among different alternatives in order to indicate whether there is a justification for intensive practices. Subsequent to this, the net benefit, which is the total benefits minus total costs, would have to be greater than in other alternatives that achieve the same flood reduction.

Flood damage quantification

The damage caused by floods is a function of the flood characteristics, i.e. depth and duration of flood inundation, due to physical contact with floodwater per category of element at risk. The flood damage estimation has therefore generally been considered to facilitate the appraisal of flood mitigation measures. In this study, the damage potential was assessed on the basis of the calculated flood depth with a 1% annual probability of exceedance for riverine flood events in order to evaluate the vulnerability to inundation, and to demonstrate the spatial distribution of potential damage across the Chi River Basin. As a result, the concerned values to elements of flood risk were calculated in order to estimate the benefits of flood mitigation measures in terms of flood damage reduction, while other impacts related to human health or environmental damage were not considered in this study.

Spatial analysis techniques, using GIS, enable integration of flood depth and land use to evaluate which elements or assets are affected by the flood depth, and how much they are affected in terms of inundation depth and duration. The following land use categories were considered in the damage assessment: residential, commercial, industrial, agriculture and infrastructure (note: damage to institutional area, i.e. government offices, was considered as part of the commercial area).

The damage functions developed by Sahasakmontri (1989) were adopted for the quantification of different damage categories in monetary terms. They provide information about the susceptibility of elements exposed to flooding. The set of damage functions are absolute damage functions in Thai Baht and expressed as given by Equation 6.4. As a result, the absolute amount of damage for each land use type as a function of the magnitude of a given inundation depth and duration has been calculated.

$$DPE = a_0 + a_1 H + a_2 L \tag{6.4}$$

where:
DPE = direct flood damage per land use type (Thai Baht)
H = maximum flood depth (cm)
L = flood duration (day)
a_0, a_1, a_2 = flood damage coefficients (see Table 6.6)

Four functions for the four categories including the coefficients for the damage functions are given in Table 6.6. From the results of a regression analysis for damage cost made by Sahasakmontri (1989), the coefficient 'a_2' in Table 6.6 has been determined to be equal to zero for the cases of commercial and agriculture. This indicates that increasing in such a long flood duration will not significantly increase direct flood damage, and only flood depth is found to be the significant variable affecting damage to both land use categories.

Table 6.6 Estimated flood damage coefficients for different land use types (Lekuthai and Vongvisessomjai, 2001)

Type of land use	a_0	a_1	a_2
Residential	-300.00	45.40	33.80
Commercial	-2.15	88.10	0.00
Industrial	-1,740.00	522.00	181.00
Agriculture	-1,050.00	553.00	0.00

Based on land use, asset values and damage functions, the direct damage caused by the flood with a 1% annual probability of exceedance was calculated as follows (Lekuthai and Vongvisessomjai, 2001):

$$DAM = \sum_{i=1}^{n} \sum_{j=1}^{4} \frac{DPE(j,H,L)}{APE(j)} \cdot PC(i,j) \cdot AREA(i) \tag{6.5}$$

where:
DAM = direct flood damage (Thai Baht)
$DPE\ (j,H,L)$ = direct flood damage per land use type j at H, L (Thai Baht /land use type)
$APE\ (j)$ = average area per land use type j per unit (m^2)
$PC\ (i,j)$ = percentage of land use type j in cell i (-)
$AREA\ (i)$ = area of cell i (m^2)
i = number of cell (-)
j = land use type 1, 2, 3, 4 (residential, commercial, industrial, agriculture) (-)
H = maximum flood depth (m)
L = flood duration (day)

At first, the flood damage was calculated using Equation 6.4 as damage per unit for each land use type. The total areas of each land use type presented in the pixel were then determined by multiplying the percentage of each land use category in each pixel by the pixel size. The number of units in each land use type was carried out by dividing the areas of the same land use for each pixel by the average area per land use type per unit in Table 6.7. The calculations were applied iteratively for four different land use categories by multiplying direct flood damage per land use type with the number of units. Thereafter, the direct flood damage calculations were obtained from the summation on a pixel-by-pixel basis.

Table 6.7 Average area per land use type per unit (after Ansusinha (1989))

	Residential (household)	Commercial (shop)	Industrial (factory)	Agriculture (farm)
Average area per unit (ha)	0.100	0.025	0.400	4.200

However, according to Sahasakmontri (1989), direct damage to infrastructure was not taken into consideration. Therefore, in this study, the damage to infrastructure has been estimated as a fixed 65% fraction of the total damage of all flood losses as estimated by Munich Reinsurance Company (1998). Using such damage functions, flood damage to different land use categories is estimated and the summation provides the total direct flood damage as shown in Equation 6.6.

$$TDM = DAM + (0.65 \cdot DAM) \quad (6.6)$$

where:

TDM = total direct flood damage (Thai Baht)
DAM = direct flood damage (Thai Baht)

Cost estimate for implementation of flood mitigation works

From the corresponding cost of flood damages, the finding whether the flood mitigation scheme is justified, cannot be taken solely on the basis of minimization of the specific costs. It is therefore necessary to estimate the implementation costs.

Estimated implementation costs for each alternative presented throughout this study are at planning level and derived from the sum of the costs associated with construction, operation and maintenance. The construction costs presented herein were determined using the handbook from the Thai Bureau of the Budget, based on April 2009 unit rates. In this regard, the annual costs of operation and maintenance were estimated by using 5% of the construction cost in order to serve the primary purposes of interventions during their designed lifespan, which typically assumes 50 years. The potential flood damage cost with a given probability of exceedance will be estimated over the 50-years project lifetime). However, the cost to implement the action is sometimes quite complicated and fraught with difficulty. For instance, in case of the costs associated with restoration of land as resulting from the occupancy of temporary or permanent flooding, some further attempt will have to be made.

The estimates presented here are subject to refinement when a more detailed design investigation is undertaken as it has not yet covered the cost of land acquisition where envisaged flood mitigation measures are to be located. Therefore, the costs presented would have to be considered as indicative only and detailed investigations are necessary to obtain more accurate cost estimates as these are probably higher in reality.

At this stage, the detailed cost estimate would have to be compiled separately for each measure according to the detailed design. In any case in which the alternatives are promising enough to proceed further, a detailed cost-benefit consideration will then have to be carried out.

Optimal selection of flood mitigation practices

The proper answer to the question 'how to choose a viable and effective alternative to achieve an optimal balance for sustainable flood management?' is the presence of an optimum of maximum benefits and minimum action costs. In consistence with Krutilla (1966), quoted by du Plessis and Viljoen (1999), as the disastrous effects of flooding continue to intensify, it is not viable or acceptable to implement measures to such an extent that they will prevent the total risk of flood losses, because it would appear that the cost of the flood mitigation measures will exceed their benefits.

To derive an optimal alternative for mitigation, each alternative needs to be evaluated individually in order to meet established levels of acceptable risk. If one of the options proves a success in prospective, it most probably is an option that will hardly be doubted and avoid the most controversial dissent. Initiatives focus on the minimum total costs of flooding (Figure 6.12), i.e. the cost of the mitigation option plus the residual damages incurred as a result of corrective action taken. In a relationship, the dotted line is the expected average annual damage cost, which is many times larger than average annual implementation cost (dashed line), however, since the implementation cost increases, the damage cost decreases. In all this, the solid line is the total cost, which is the sum of those two costs.

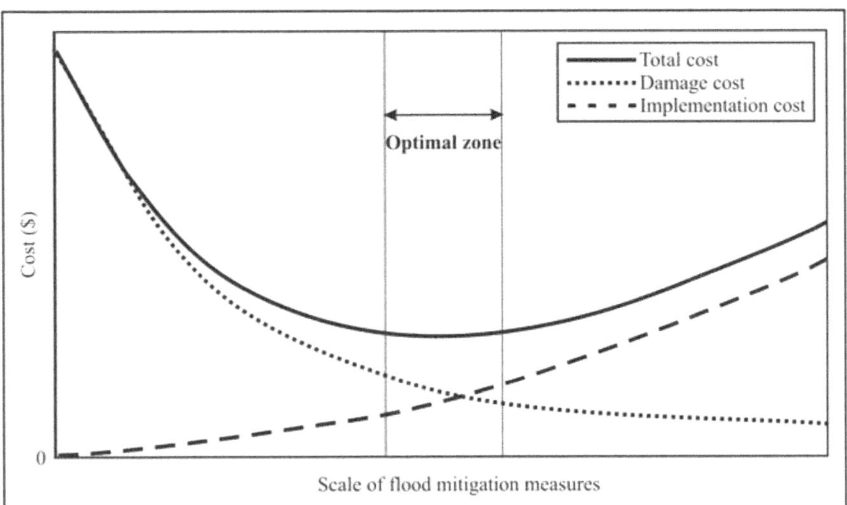

Figure 6.12 Financial optimum as defined by cost benefit analysis (adapted from Woodall and Lund (2009))

6.3.3 Determination and appraisal of promising flood mitigation measures

Prior to envisaging the selection of possible mitigation efforts, the decision cannot be taken based on a single indicator, which is the minimization of the cost. It is therefore necessary to involve more than one criterion. As a result, more technical effectiveness criteria associated with each scenario are then examined. To make this happen, the 1D/2D SOBEK hydraulic model has been used to determine areas, which might be prone to flooding from a flood with a 1% annual probability of exceedance. The model has also been used to identify areas where the proposed options might be employed and will not cause adverse impacts on existing flooding conditions downstream. To understand the impact of such potential measures in flood damage reduction, selected scenario analysis is therefore required in order to choose the preferable alternative.

Efficiency of alternative flood mitigation measures

Theoretically, it is possible to reduce potential flood damage, if the right selection and implementation of various mitigation techniques and measures is made. In hydraulic systems, it is typically necessary to evaluate their hydraulic effectiveness on a common basis as some measures would affect the flood behaviour and potentially exacerbate the flood risks in some areas. Therefore, there is a need to compare the flood characteristics, i.e. flood depth and flood extent, to which flooding would be lessened under the situation with and without the flood alleviation measure concerned. To assess the variation in flood behaviour, the relevance of each of these measures will be briefly described and considered to strengthen flood sustainability in the Chi River Basin.

Decision matrix for alternative flood mitigation measures

Flood management interventions involve a number of alternative structural and non-structural measures that need to be assessed and quantified. However, only the most common measures recently recommended that can strengthen an integrated approach to flood management will be discussed. Hence, a decision matrix approach will be used to guide which flood mitigation measures would be considered in this study. At first, the steps are defined by generating a range of measures, assessing the expected performance of each measure against the evaluation criteria, and selecting the preferred options. The alternative flood mitigation measures will be considered, only if they can meet the evaluation criteria. Once these criteria have been applied, selected measures will be discussed to which alternatives are the most applicable and desirable. The following evaluation criteria have been used to screen and prioritize proposed flood mitigation measures:

- *feasibility*:
    - is it expensive?
    - is it cost-beneficial?
    - is there any funding available (capacity to implement and manage the project)?
- *technical effectiveness*:
    - will it work?
    - does it solve the problem (potential flood damage reduction)?
    - is it feasible?
    - is it built to a certain safety level that can be exceeded by a larger flood or by overtopping or failure of the structure, causing even more damage than might have occurred without the structure?
    - can it lead to a false sense of security as people protected by a structure believe that no flood can ever reach them?
    - does it require regular maintenance to ensure that it continue to function properly and provide its design safety level?

- can it alter the timing of flood peaks and divert floodwater onto other adjacent unprotected properties, which will potentially reduce the floodplain storage capacity?
- is it compatible and flexible with other flood mitigation measures?
- can it adapt itself readily for other purposes/new functions?
- does it promote more intensive land use and development in the floodplain?
- can it be built with minimal disruption of homes and businesses?

Each flood mitigation measure has been scored with respect to a set of indicators for each evaluation criterion by conducting a review of performance in order to assess and prioritize flood mitigation measures. This is more easily explained using the sample score table. The indicators were chosen to represent criteria important for deciding which flood mitigation measure ultimately best meets the overall objective.

In setting up the priorities, preference is given, first of all, to the flood mitigation measures able to reduce flood risk and damage in correspondence with designated evaluation criteria. For each of the indicators, score categories have been described qualitatively, which reflect the relative attention to each flood mitigation measure, including the absolute amount or investment. Then, for each indicator a score is assigned to each flood mitigation measure. The less the star marks (*) is assigned, the more the measure is unfavourable, in contrast more star marks means the measure is more favourable.

6.3.4 Potential flood management opportunities

With the hardship and devastation caused by floods, it ought to seriously address both identified problems and opportunities related to flood management. As a matter of fact, the bigger the flood, the less potential flood mitigation can feasibly lessen the consequences, regardless of the mitigation type. Therefore, in order to tackle particular flooding problems, it is associated with more than just defending against floods.

Some evidences suggest that different flood management options previously appeared impracticable and technically unachievable. Therefore, a range of potential flood mitigation options is herein presented in response to a challenging query, i.e. how opportunities are being exploited for a flood management system to deliver more flexible and cost-effective solutions. Complying with this principle seems straightforward, but it can affect the physical balance and would imply that additional careful consideration is required, i.e. to highlight the variety of ways and is not always tightly involved to engineering solutions. In an attempt to clarify the potential flood mitigation opportunities, each measure, both structural and non-structural, is therefore evaluated with respect to its primary potential gains and constraints.

Towards simulation of inundation based on a flood disaster scenario with the annual probability of exceedance of 1%, there have been four possible flood mitigation alternatives analysed in order to estimate their hydraulic effectiveness in reducing floods in critical sites of the Chi River Basin. Potential hydraulic impacts of flood mitigation were assessed by comparing the results of pre- and post-mitigation extents of flood depth and inundation. These are: river normalisation, reservoir operation, green river (bypass channel) and retention basin.

River normalisation

Flood disasters are caused by discharge that is higher than the capacity of the existing river causing water to overflow to the adjacent lands. The river capacity is often reduced because of the narrowing due to extensive vegetative growth, accumulation of drift and debris, or blockage by leaning or uprooted trees. This is the reason why river normalisation can be a viable option; in some cases it can be done by reducing hydraulic roughness, increasing cross-sectional area, or reducing potential for blockages and hang-ups of drift. For that reason, careful attention would have to be paid to the selection of methods to improve the hydraulic characteristics of the river, which should not disrupt the existing morphological balance of the river system. To compensate for the alteration in the hydraulic variable and to establish a new/stable equilibrium, other parameters will change. Therefore, river improvement will probably have a great impact on a river because it can disrupt the existing physical equilibrium of the river system unless the operations are carefully planned, conducted, and monitored. Furthermore, regular maintenance would also need to be carried out to ensure continued satisfactory operation.

In this study, river normalisation works are proposed to increase the velocity of the river flow and consequently lower the flood stage through the Chi River by cleaning a river channel. It may include the removal of logs, drifts, snags, boulders, piling, debris, and other obstructions from the flow area of river. As a result, the inundation of the adjoining land can be decreased. However, it may also have an effect of increasing the peak discharge of the flood, which will inevitably create worse flood conditions downstream over that before river normalisation.

The estimation will be made in order to identify potential effects of river normalisation on stream function, which will change the flood runoff of the river. Therefore, river normalisation is then modelled by changing the hydraulic variable, i.e. river roughness, in the routing model. Manning's roughness coefficient as shown in Equation 6.7 has been applied (Chow et al., 1988).

$$v = \frac{R^{2/3} s^{1/2}}{n} \tag{6.7}$$

where:

v = mean flow velocity (m/s)
R = hydraulic radius (m)
s = slope of energy grade line (-)
n = Manning's roughness coefficient (s/m$^{1/3}$)

The calibrated Manning's roughness coefficients were reduced to reflect the changes that occur during the process of river normalisation. To illustrate this alternative, the river after the completion of river normalisation is expected to have a clear river channel. However, estimating Manning's roughness coefficients for river normalisation work is a matter of judgment and shall only be applied in reaches that yield the highest net benefits. In this case, the Manning's roughness coefficient for reaches to be modified by river normalisation was arbitrarily set to 0.025. As set out in Table 6.8, the suggested range of Manning's roughness coefficients for selected normalisation reaches were defined for use in the simulations.

Referring to Figure 6.6 and Table 6.8, in Scenario RN_I, the physical normalisation was proposed to be done mainly from the upstream gauged site (E5) through the Chi-Nam Phong confluence (reach 1, 3, 4 and 5). A further intensive normalisation was continuously assigned until the Chi-Lam Pao confluence (reach 1, 3, 4, 5 and 6) (Scenario RN_II), while the normalised active channel in Scenario RN_III was supposed to be expanded until it reaches the Chi-Nam Yang confluence (reach 1, 3, 4,

5, 6, 7 and 8). As for Scenario RN_IV, the dynamic potential of the normalised river was specified over an entire river reach. The idea of normalisation of the downstream river segment (Scenario RN_V) was also presented and carried out between the Chi-Nam Phong confluence and the Chi-Mun confluence (reach 6, 7, 8 and 9). The normalisation work from the Chi-Lam Pao confluence through the Chi-Mun confluence (reach 7, 8 and 9) was addressed in Scenario RN_VI, while Scenario RN_VII was only suggested at the far end of the Chi River (reach 9: from the Chi-Nam Yang confluence through the Chi-Mun confluence). In this regard, it is of interest to note that no action has to be taken on any reach that was not mentioned in each scenario.

Table 6.8 Manning's roughness coefficients corresponding to different normalisation locations

Reach	Calibrated Manning's n	Scenario						
		RN_I	RN_II	RN_III	RN_IV	RN_V	RN_VI	RN_VII
1,3,4,5	0.045	0.025	0.025	0.025	0.025	0.045	0.045	0.045
6	0.032	0.032	0.025	0.025	0.025	0.025	0.032	0.032
7,8	0.032	0.032	0.032	0.025	0.025	0.025	0.025	0.032
9	0.028	0.028	0.028	0.028	0.025	0.025	0.025	0.025

The following involves the characterisation of the flooding scenarios, identifying the consequences of the scenarios and evaluation of flood reduction measures.

From an financial view point, the costs of river normalisation, which can fluctuate significantly based on the length of normalisation, are estimated at about 18,000 US$/km (see Table 6.9 for details).

Table 6.9 Financial evaluation of river normalisation alternative for scenario screening (unit cost = 18,000 US$/km)

Scenario	Length of normalisation in km	Cost in million US$			
		Implementation		Damage	Total
		Construction	O&M		
No measure	-	-	-	86	86
RN_I	360	7	16	84	107
RN_II	550	10	25	80	115
RN_III	660	12	30	74	116
RN_IV	880	16	40	62	117
RN_V	520	9	23	64	97
RN_VI	330	6	15	64	85
RN_VII	220	4	10	73	87

Note: O&M = Operation and maintenance

In evaluating different techniques, the flood mitigation effects of the particular set of scenarios were estimated in a quantitative manner based on the hydraulic modelling. The effect of flood mitigation alternatives on potential adverse consequences is graphically depicted in Figure 6.13 by specifically referring to the effectiveness of mitigation and adaptation measures. As expected, almost all scenarios produces significant benefits (Figures 6.13a and 6.13b), however, it is still difficult to judge which scenario will prevail. Therefore, a cost-benefit analysis that leads to minimum total cost has become a valuable tool to highlight the differences between individual scenarios.

As illustrated by Figure 6.13a, when the river normalisation works are mainly implemented downstream of the Chi River (Scenario RN_V, VI, and VII), the shallow flood depth (0 - 1 m) at the upstream part will increase. Since the upstream reaches have not been implemented, the water within a river channel cannot flow fast enough which causes an overflow onto the floodplain, where the flood depth starts rising up between 0 - 1 m. On the other hand, the water can flow faster at the downstream reaches, which have been assigned to be normalised, As a result, the flood depth on the floodplain tends to decrease in the downstream part.

A meaningful comparison of scenarios in Figure 6.13c suggests that all scenarios other than Scenario RN_VI are objectively judged as implausible and unlikely to be qualified. Although some of them provide a significant reduction in flooding (Scenario RN_IV and V), but they fail to achieve the optimum cost benefit possible or minimum total cost (Table 6.9). If properly performed in the form of normalising the reaches from the Chi-Lam Pao confluence through the Chi-Mun confluence (Scenario RN_VI), the flood characteristics are found to be significantly reduced. Changing the Manning's roughness coefficient from 0.032 (between the Chi-Lam Pao confluence through the Chi-Nam Yang confluence) and 0.028 (between the Chi-Nam Yang confluence through the Chi-Mun confluence), to 0.025 causes the extent of 1% annual probability of exceedance flood to decrease around 16,000 ha.

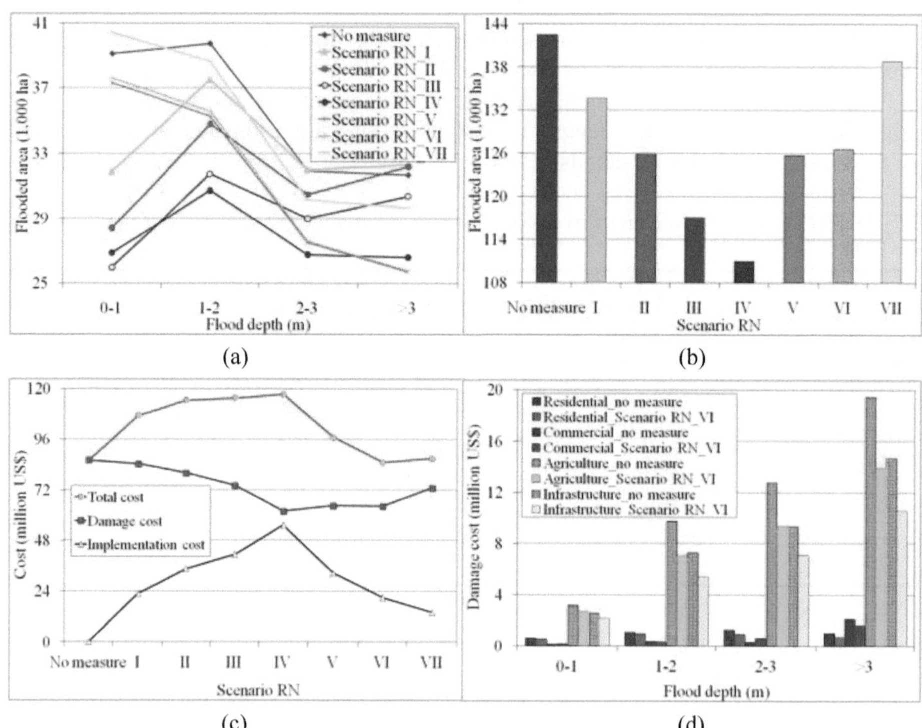

Figure 6.13 Four types of illustration of results from flood modelling under river normalisation alternative: a) flooded area at each flood depth for all scenarios; b) flooded area under different scenarios; c) cost-benefit analysis of the scenarios; and d) damage cost per land use categories under the optimal scenario

Towards a comparison between with and without Scenario RN_VI, the cost of damages associated with flooding is expected to decrease from US$ 86 million to US$ 64 million across the river basin (Figure 6.13c and Table 6.9). In Figure 6.13d, it is found that the potential loss from flood damage under Scenario RN_VI is mainly contributed by the agricultural sector, which represents approximately 52% of the total damage. As can be seen, Scenario RN_VI reduces the impacts, but does not completely eliminate them. However, this scenario is still preferable and recommended as its benefits are in excess of the implementation cost, with benefits of US$ 22 million compared to estimated costs of US$ 21 million.

Reservoir operation

As noted by its name, the interest starts with emphasis on reservoir management to avoid needless and excessive flood damage. In any sense, whatever the underlying cause may be, an immediate and sensible solution needs to be taken to rectify the situation. To that extent, flood storage reservoirs can lead to better regulation of the flow regime and therefore minimize potential flood damage. However, the land to form a new surface reservoir to capture floodwaters along a river and temporarily store water in the reservoir are currently no longer available, due to the massive displacement of people in the last few decades.

To assure the degree of safety to the flood prone areas, modification of the reservoir operation rule is considered to reduce economic disruption caused by floods, as well as livelihood and well-being of riverine and floodplain dwellers. The reservoir release patterns following the operation rule curve were used to cope with all magnitudes of flood events. However, this operation might not be executed with some flood events due to the increase in the complexity of reservoir operation, i.e. conflicting reservoir purposes.

Presently, the flood control potential of the existing reservoirs, i.e. Ubol Ratana and Lam Pao is limited because of the larger volumes of flood runoff generated in the Chi River Basin (Figure 6.6). To maximize the flood alleviation benefits downstream, the reservoir operation rule curve would have to be modified whereby the reservoir level is deliberately lowered during the rainy season to store flood runoff. The way a reservoir is operated can reduce or attenuate the peak discharge, delay or lag the timing of the flood peak and modify flood releases downstream of the reservoirs. In addition, the amount of storage required depends upon the degree of protection needed and the downstream channel capacity. Therefore, the way in which reservoir controls are operated can affect the degree of downstream mitigation. The reduction of the flood peak is at the expense of:

- releasing part of the water of those two reservoirs to the irrigation schemes prior to the rainy season, in association with the redesign of the cropping pattern. In other words, cropping calendars would have to be adapted to the expected time of year, and frequency and severity of flooding;
- prolonging the duration of reservoir releases at lower rates during the rainy season.

The main objective of this scenario is to present the various options of reservoir operation, and the possibilities of how flows need to be managed to minimize their adverse impacts while also maximize their beneficial impacts. Five scenarios of reservoir operation described below were compared to investigate the effectiveness on reduction of flooded areas (Table 6.10).

Table 6.10 Percentage of daily outflow and initial reservoir volume of each scenario for Ubol Ratana and Lam Pao reservoirs

Reservoir	Scenario									
	RO_I		RO_II		RO_III		RO_IV		RO_V	
	Rel	Vol	Rel	Vol	Rel	Vol	Rel	Vol	Rel	Vol
Ubol Ratana	90	90	75	75	55	50	100	100	55	50
Lam Pao	90	90	85	60	75	30	75	30	100	100

Note: Rel = Daily reservoir outflow in % of the original daily outflow
Vol = Initial reservoir volume in % of the original initial volume

Due to the fact that the performance of a reservoir depends on its capacity, configuration, location, and operation rules, the reduction in initial reservoir volumes and release operations are considered important and therefore included as a factor to reduce the impacts of a 1% annual probability of exceedance flood event. The percentage of reservoir outflow considered the inflow, outflow, losses, and reservoir storage with the reservoir water balance Equation 6.8, which was determined by the general hydrology of the river at the reservoir site and over its entire surface, as well as the reservoir operation for achieving its basic functions. The reservoir releases have also been determined by taking into account the downstream needs for water supply, agriculture, and environment, by requirements for flood regulation, storage considerations, and legal requirements for minimum flows.

$$S(t+\Delta t) = S(t) + (I \cdot \Delta t) - (O \cdot \Delta t) + (P \cdot \Delta t \cdot A \cdot 10^3) - (E \cdot \Delta t \cdot A \cdot 10^3) \quad (6.8)$$

where:

S = water volume V (m^3) stored in the reservoir ($0 \leq S \leq V_{max}$) (m^3)
I = tributary inflow to the reservoir (m^3/day)
O = outflow from the reservoir (m^3/day)
P = rainfall on the reservoir (mm/day)
E = reservoir evaporation and other losses (mm/day)
A = reservoir surface area (km^2)
t = time (day)

In the above mentioned scenarios, it appears that no preliminary view can be set forth because each scenario is unique and needs to be evaluated. Thereby, initial attempt to evaluate the success of this operation scheme is necessary before any decision can be made. Prior to this, it is important to point out here the details of each scenario, which are given in the following. The scenarios with particular regard to the variations in daily outflow and initial reservoir volume at the beginning of the wet monsoon for Ubol Ratana (UB) and Lam Pao (LP) reservoirs were defined to evaluate the sensitivity of flood damage estimations to these changes (Table 6.10 and Figure 6.14b).

- for Scenario RO_I, about 90% of its original daily outflow would be released from UB and also 90% from LP during the wet season. Taken together, their initial volumes would be approximately 90% (UB) and 90% (LP) of their original initial volumes;
- Scenario RO_II involved the daily outflows at approximately 75% (UB) and 85% (LP) of their original rates, while the initial volumes of both reservoirs were prescribed at 75% (UB) and 60% (LP) of their original initial capacities;

- the outflows compared to their original releases in Scenario RO_III were assumed to be 55% (UB) and 75% (LP), in which the percentages of initial volume were assigned to be 50% (UB) and 30% (LP) of their original initial volumes;
- Scenario RO_IV kept the same outflow rate for UB but reduced from its original daily outflow for LP at about 75%, together with no change in UB initial storage and 30% of its original initial capacity applied to LP;
- in Scenario RO_V, the outflow from UB was limited to 55% of its original daily outflow and there was no change in outflow for LP, whereas UB initial storage capacity was reduced to 50% of its original initial capacity and remained the same for LP.

It is noted that both original daily outflow and initial reservoir volume are based on the reservoir operating rule information during the period from June - December 2001, and take into account the effects of maximum 3-day rainfall with a 1% annual probability of exceedance.

The results revealed that Scenario RO_III is the best option due to the low inundated area, which represents the ability to control flood peaks as compared to other alternatives (Figures 6.14a to 6.14c). By following the proposed reservoir operation rules, the flooded areas can be reduced by 10% as illustrated in Figures 6.14b and 6.14c, while both reservoirs can significantly increase the amount of storage space to store impending floods up to their storage capacity until the end of the flood season.

Regarding the drawing up of an analysis of the costs and benefits, the scenario that yields the highest effect has been determined. Under this scheme, the implementation costs were considered to be zero as change in the operation rule curves for the reservoirs would not require any construction, therefore no additional expense is anticipated. Within the criterion concerned, Scenario RO_III is also considered more attractive than others largely due to the reduction in the value of damage for a flood with a 1% annual probability of exceedance by US$ 16 million or 19% (Figure 6.14c and Table 6.11). By relating the results to the benefits for this scenario, this action would reduce the flood damage potential for each land use type, especially with greater decrease in agricultural damage at higher flood depth (Figure 6.14d). To this end, it can be noticed that this option is concrete enough to be sufficiently plausible and meaningful since it provides similar benefits as the river normalisation option, although it results in a slightly larger flooded area.

However, operation of reservoirs needs very careful consideration in further studies, with special attention paid to direct and indirect damages, including reduction in benefits from the reservoir in terms of hydropower generation, irrigation, water supply, etc., which may make the reservoir operation alternative expensive. There is also a perception that substantial reduction in initial reservoir storage to accommodate incoming flood of significant magnitude, is likely to cause potential drought-related impacts and risk next year. Nevertheless, one should bear in mind that the overestimation of the initial reservoir volume and reservoir outflows can lead to substantial losses. Therefore, timely and reliable reservoir inflow forecasting for initial reservoir storage during the rainy season could significantly enhance the accuracy and efficiency of reservoir operation guidance. In addition, reservoir operation alone cannot be considered as a reliable alternative to safely pass tremendous floods. Therefore, it would be more trustworthy if this alternative would have to be integrated with other flood mitigation measures, e.g. through legislation and controlling the developments in the floodplains.

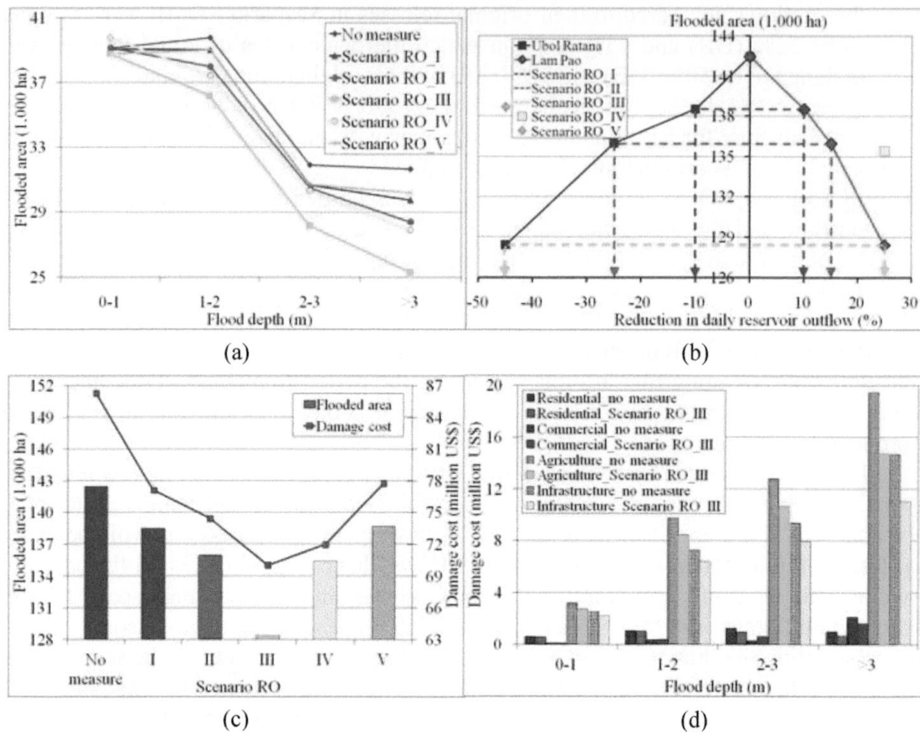

Figure 6.14 Four types of illustration of results from flood modelling under reservoir operation alternative: a) flooded area at each flood depth for all scenarios; b) flooded area under different scenarios; c) cost-benefit analysis of the scenarios; and d) damage cost per land use categories under the optimal scenario

Table 6.11 Financial evaluation of reservoir operation alternative for scenario screening (no implementation cost for this alternative)

Scenario	Cost in million US$	
	Damage	Total
No measure	86	86
RO_I	77	77
RO_II	74	74
RO_III	70	70
RO_IV	72	72
RO_V	78	78

Green river (bypass channel) measure

In the Chi River Basin, the floodplain is always flooded by the floods travelling down from streamflow station Ban Muang Lad (E66A) through the confluence of the Mun River (Figure 6.15). Under most circumstances, flooding takes place through overtopping of low-lying points on the river banks of the Chi River.

To mitigate flooding problems by intercepting and controlling peak discharges upstream of a flood prone area, a 'green river' is proposed to divert the excess floodwaters from the Chi River through the bypass channel, in order to keep the flow within the capacity of the Chi River. This will decrease the impacts of damaging floods in the river system as a whole. In this instance, an extra outlet would be connected thereto for water from upstream to improve flow characteristics by decreasing levels of water and slowing down the release of water for some distance downstream from the point of diversion. To do this, such channel will act as a flood detention basin to temporarily store a portion of the floodwater and thus lessen the flow in the Chi River.

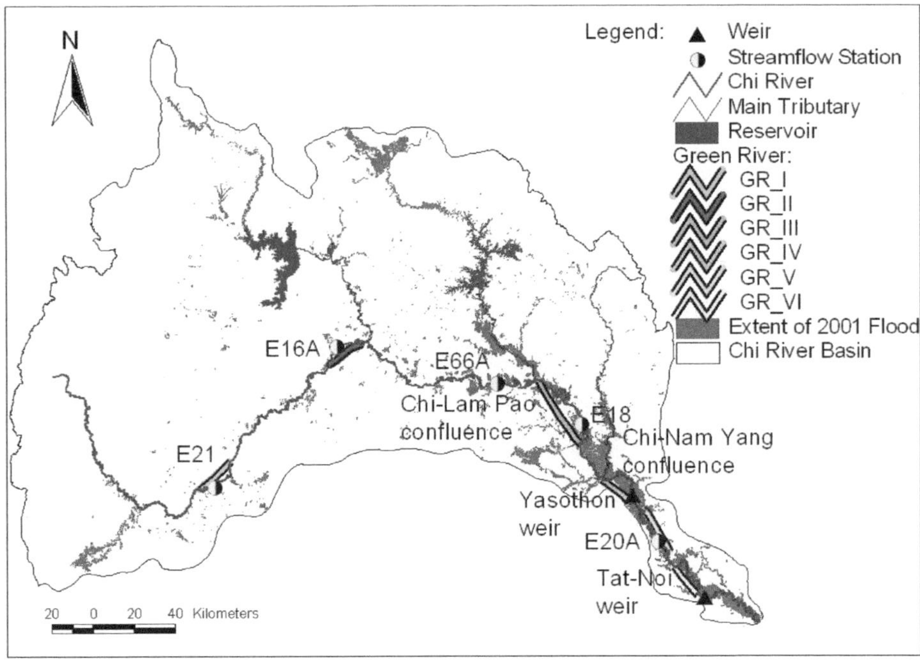

Figure 6.15 Designated green river channels along the Chi River

The green river measure can be effective, if the following conditions can be met:

- free from human settlements;
- bordered by embankments to confine the floodwaters;
- having reasonably effective drainage facilities.

Moreover, it is also recognized that the potential benefits of using green river channels for flood damage mitigation include the following:

- green river channels are only used during major floods, so they can be used as water retention for dry season irrigation. In addition, since it may be dry during non-flood times, it is often desirable and may be accepted to use land in this area for agriculture, wetland, greenbelt, pasture for grazing, or wildlife food source purposes;
- existing developments in flood prone areas can be protected from frequent flood damage without expensive internal drainage alterations;
- it is environmentally beneficial to use green river channels as an alternative to modify the Chi River to convey flood flows, because the periodic flood conditions can favour the extent of potential for growth of riparian vegetation;
- in combination with other flood modification methods, diversions can be tailored to site-specific flooding problems.

However, there are some limitations with the green river measure, which may comprise the following:

- its construction opportunities are sometimes limited by unfavourable topographic conditions;
- non-availability of low-value land, which can be used for this measure;
- in urban areas where land is scarce, this may not be a viable option because the land cannot be developed;
- dwellers may occupy the land exposing them to increased risk when the green river channels are in use;
- any alteration to the natural flow regimes of the main river involves some changes in hydrology with possible negative impacts;
- the diversion system must be carefully designed and constructed to prevent channel instability in the main channel and the diversion channel;
- if adjacent low-lying areas are used for diversion purposes, some terrestrial habitat may be lost or converted to a wetland habitat;
- a diversion can have a high capital cost, public funds are limited and a false sense of security against flood damage may develop.

In this alternative, green rivers will be applied to determine whether they can improve flood flow capabilities in the flow passage through its downstream end as illustrated in Figure 6.16 the expected impact of green river channels on adjacent water. To accomplish this, efforts need to be made to ensure that the channels will function as designed. Green river channels would have to be properly designed and maintained to accommodate the peak flow and velocities for the areas susceptible to damage and subject to most frequent flooding. No development would be allowed within the designated green river channel since extensive damage may result from deeper and more swiftly flowing waters.

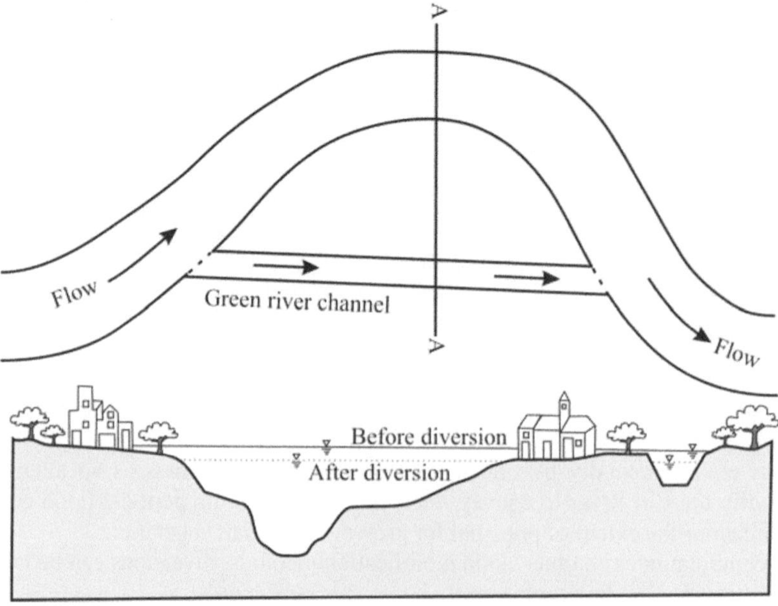

Figure 6.16 Effects of a green river channel

Under this alternative, the fact that hydraulic design is found to be complicated. Therefore, the detailed design is largely arbitrary. In light of this, opportunities for a series of green river reaches were considered across the Chi River Basin. The preliminary alignment shown in Figure 6.15 gives the possible locations where potential alignments to shorten the river course occur. For effectively diminishing the threat of flooding, the green river channel was designed as a trapezoidal channel of 25 m width at the base (top width of 31 m) with the design channel side slope 1H to 1V and 3 m height, as illustrated in Figure 6.17. The enlarged channel would allow flood flows to start bypassing at an earlier stage than currently, with increasing flows taken by the green river as the river rises. The control is in the form of spillways or gates at the entrance to a green river.

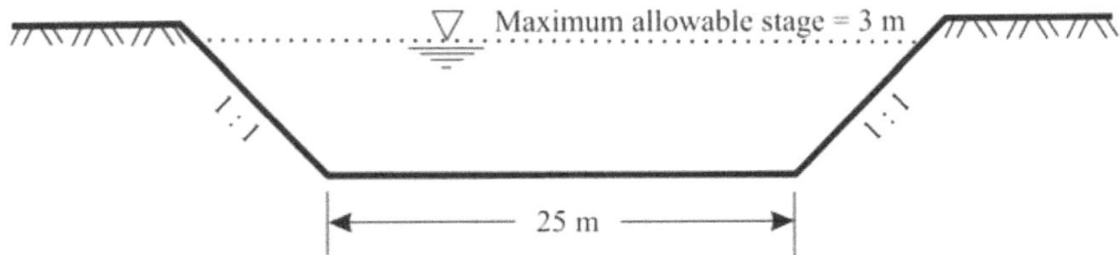

Figure 6.17 Cross section of green river channel

The flood discharge with a 1% annual probability of exceedance will be diverted into the green river channel and safely conveyed to the Chi River. As a result, floods fill the channel and rise within the green river channel, after some time, the flow is released back into the Chi River at a distance downstream from the point of diversion. With a defined starting point, a sufficient enough distance between the point of diversion and point of re-entering the Chi River is required to avoid backwater effects. The proposed green river channel with a proper length was therefore defined in parallel with the Chi River comprising of six local drainage channels with preliminary dimensions as shown in Table 6.12. In addition, a potentially favourable green river channel option has been considered as being important for flood mitigation (Kuntiyawichai et al., 2010a; Kuntiyawichai et al., 2010b; Kuntiyawichai et al., 2011a; Kuntiyawichai et al., 2011b). However, this option is probably feasible in some specific areas in the Chi River Basin. Hence, it will require assessment on a site-by-site basis in order to determine its effects upon the flood characteristics. Therefore, every scenario would need to be tracked and elaborated the details of what the scenarios are about (Figure 6.15).

- as for bypass in Scenario GR_I, about 20 km in length and situated on the left bank floodplain of the Chi River, it is aligned to divert floodwater from a distance of 16 km upstream of Ban Kaeng Ko (E21) gauged site to a designated point 30 km downstream of the same gauged site;
- an 18 km long diversion channel on the right bank floodplain of the Chi River (Scenario GR_II) was considered to carry floodwater off the point 19 km upstream of Ban Tha Phra (E16A) streamflow station and put it back at 22 km downstream of the same gauging station;
- the notion of Scenario GR_III was addressed with regard to floodway on the right bank floodplain of the Chi River. It stretches approximately 36 km to direct excess water from the vicinity of Chi-Lam Pao confluence to the junction 22 km downstream of Ban Tha Khrai gauged site (E18);

- Scenario GR_IV involved a 16 km flood relief channel, which lies on the right bank floodplain of the Chi River and is assigned to divert floodwater from the Chi-Nam Yang confluence to the meeting point 6 km downstream of Yasothon weir;
- the channel on the left bank floodplain of the Chi River under Scenario GR_V diverts floodwater from 21 km downstream of Yasothon weir through a 23 km channel into the release point at 11 km downstream of Maha Chana Chai (E20A) gauged site;
- Scenario GR_VI was characterized by a 17 km long relief channel on the right bank floodplain leaving the Chi River at 27 km downstream of Maha Chana Chai (E20A) gauged site and rejoining the river at 5 km upstream of Tat-Noi weir;
- in Scenario GR_III+VI, it assumed a combination of Scenario GR_III and VI.

To be clear, the distance, which was used to indicate the points of divergence and convergence of the green river channels from the reference points such as gauged sites, confluences and weirs, in this alternative referred to the main channel distance.

The findings regarding the relationship between flood depth and flooded area is likely to have a significant impact (Figure 6.18). Under this set of scenarios, it seems likely that Scenario GR_III will reduce the maximum extent of flood-affected areas of shallow flood depth (less than 2 m), while Scenario GR_VI generates the least, particularly for depths greater than 2 m. Since Scenario GR_III and Scenario GR_VI are two of the best for different flood depth zones but no one optimizes, a final scenario (Scenario GR_III+VI) has therefore been created by combining them. In this case, the upstream river reach does not have to be implemented, as most of the floodwaters would now enter the green river channel three and six through gaps, which were specifically designed for that purpose. In circumstances like these, the flow can be managed with a substantial reduction in downstream flooding. Obviously, the hydraulic model demonstrates that there is a potential for a decrease in flooded area of about 8,000 ha (Figure 6.18b). The results are also supported by the fact that the damage cost could be reduced by approximately US$ 10.8 million (Figure 6.18c). In comparison, the implementation cost of the green river channel is estimated at US$ 9.3 million, whereby a net benefit of US$ 1.5 million is gained. With the obvious merits this alternative brings, it would do the most to reduce crop damage as it reveals some reduced agricultural damage compared to the baseline case (Figure 6.18d).

Apart from its versatility, it is also essential that a certain level of maintenance and upkeep of the green river channel would have to be considered so that floodwaters could pass through easily when needed. At last, to the fact that whichever route of green river channel is put in place, the potential adverse effects still exist upon downstream flood conditions. For this reason, one of the most remarkable necessity connected with the remaining consequences, is the extent of this alternative with the right mix of other options.

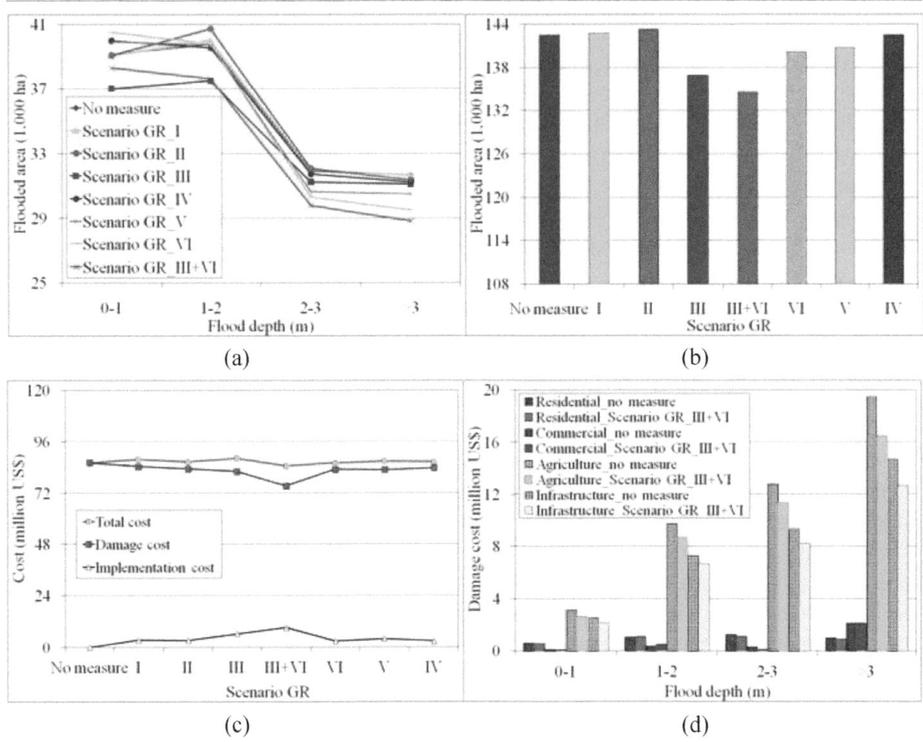

Figure 6.18 Four types of illustration of results from flood modelling under the green river channel alternative: a) flooded area at each flood depth for all scenarios; b) flooded area under different scenarios; c) cost-benefit analysis of the scenarios; and d) damage cost per land use categories under the optimal scenario

Table 6.12 Financial evaluation of green river channel alternatives for scenario screening (unit cost = 50,400 US$/km)

Scenario	Length of green river in km	Cost in million US$			
		Implementation		Damage	Total
		Construction	O&M		
No measure	-	-	-	86	86
GR_I	20	1	2	84	88
GR_II	18	1	2	83	87
GR_III	36	2	4	82	88
GR_IV	16	1	2	84	87
GR_V	23	1	3	83	87
GR_VI	17	1	2	83	86
GR_III+VI	53	3	7	75	85

Note: O&M = Operation and maintenance

Retention basins

With implementation of the necessary mitigation measures, the proper schemes will not lead to unacceptable increase in flood risk to areas elsewhere upstream or downstream of/adjacent to the development. In reality, natural storage on floodplains, which acts like (giant) sponges to soak up floodwater has been reduced by agriculture, urban developments, etc., leading to an increase in flood peaks, severity of flooding, and loss of terrestrial habitat. For the Chi River, reliance on protection dikes and on existing reservoirs can only attain a limited level of flood control. An effective way of protection against floods is a retention basin, which aims to reduce flood risk to the communities and also to alleviate flood damage in the river basin as a whole. Retention basins are flood storage areas, which perform a similar function as upstream reservoirs in storing a portion of the flood volume and diminishing the flood peak. Due to the fact that retention basins are typically located in areas which provide natural storage, hence without adequate retention basins, flood discharges and flood damages may take place at downstream locations. In doing so, specific facilities with considerable storage volume will be required if the retention basin measure is adopted. This measure allows floods greater than a specified magnitude to temporary spread over low-lying areas situated behind dikes in association with the operation of gate structures, and eventually releases it slowly back into the river after the peak has passed. In addition, since the storage is not within the conveyance system, floodwater may be stored as long as desired to achieve the necessary improvement in peak flow reduction. However, more detailed flood modelling still needs to be conducted to ensure that a new retention area will not cause adverse impacts on existing flooding conditions downstream.

It needs to be noted that randomly placed retention basins may actually increase peak flows. Therefore, the location, size and outlet for each retention basin would have to be selected on a site-specific basis depending on the storage characteristics of the basin and nature of the identified flood problem to be solved.

Before a retention basin can be designated, it is necessary to determine which areas might be prone to flooding from an event with a 1% annual probability of exceedance. Thereafter, the retention basin begins to pass the flood discharge into the storage facility when the discharge in the Chi River reaches the critical value. As a result, the retention basin fills and impounds floodwater. When the inflows subside, the retention basin continues to release the stored floodwater over an extended period of time until complete drainage. To achieve the required storage, it may be necessary to build dikes and control structures to allow particular areas of the floodplain to be inundated to specified depths at particular times. Thus, the retention area operation will be approximated by the inlet and outlet control capacity determined by gate structures. In addition, the discharge capacity of the inlet and outlet for a full retention basin needs to be equalled the maximum flow in which the flood diversion channel can pass without causing damage. The basic elements that control floodwater through a retention basin (Figure 6.19) include:

- an inlet work at the entrance to collect floodwater;
- a storage basin to detain floodwater;
- a flood diversion channel to convey floodwater to the retention basin (notation relates to Figure 6.17);
- an outlet structure to control the release of floodwater.

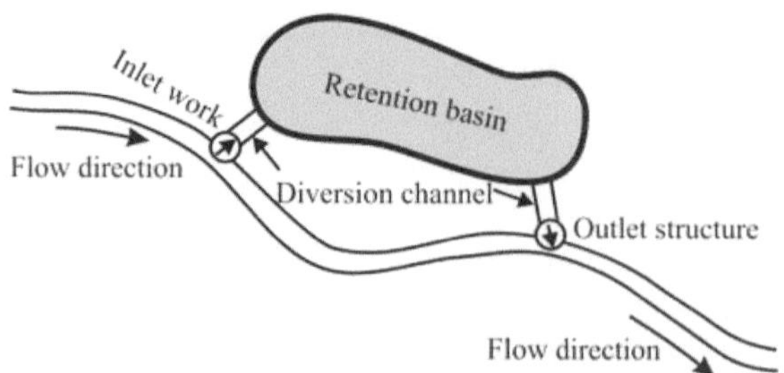

Figure 6.19 Basic components of retention basin

From the design point of view, since the retention basin requires a large area of low-lying land, thereby the area to be utilized needs a careful selection as it is not easy to find available sites. To put in place the flood retention basin, agricultural land including naturally depressed area, which is subject to repeated flood threats and could be hydraulically connected to the Chi River has been targeted as potential site to ensure that flow in the Chi River will not worsen existing flooding problems. If this alternative would be applied, the proposed site is recommended to be zoned for less intensive agricultural activities as the area will be flooded for certain periods of time. In this case, retention basins between 4,500 to 5,500 ha of estimated usable area were considered as they have to be based on the available land with the proper topography. The approximate locations of the designated retention sites are shown in Figure 6.20.

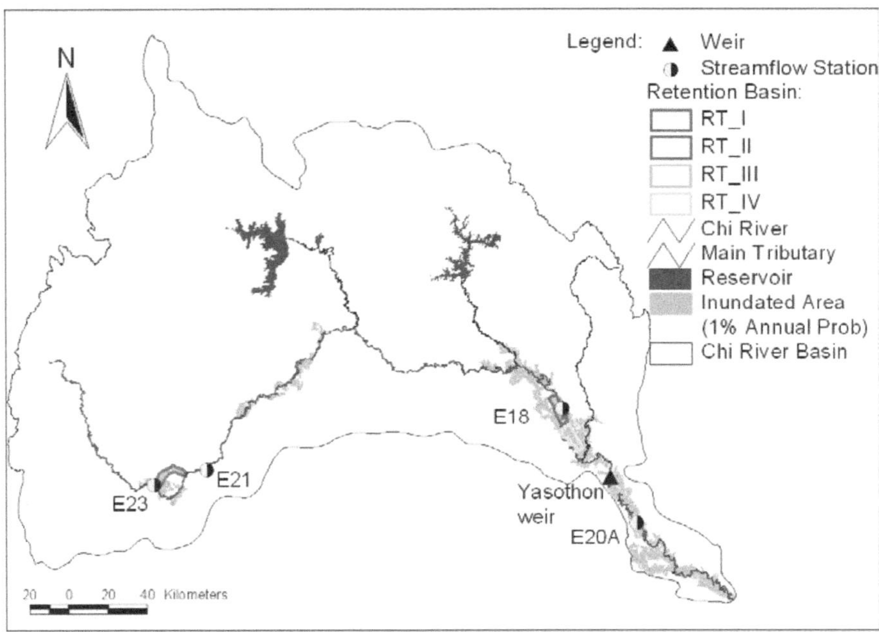

Figure 6.20 Indicative locations of proposed retention basins

Yet it might be equally hard to think about a complete picture of what these scenarios will look like without any prior explanation. Therefore, it is of importance to provide a brief description of each scenario as listed below (note: the retention basins considered herein are surrounded by earthen ring dikes).

- Scenario RT_I involved the creation of upstream flood storage on the left bank floodplain of the Chi River with a height of 9 m and a total circumference of 43 km. The usable storage volume is approximately 5 million m^3 in an area of 5,500 ha. It is undertaken between Ban Khai gauged site (E23) and approximately 22 km upstream of Ban Kaeng Ko (E21) gauged site;
- Scenario RT_II included the provision of smaller downstream flood storage space, which has a surface area of 5,000 ha with a 7 m high earth-fill dike, total perimeter of 41 km and storage capacity of 3.5 million m^3. This is situated on the right bank floodplain of the Chi River between the point 8 km upstream of Ban Tha Khrai gauged site (E18) and about 14 km downstream of the same gauged site;

- downstream retention basin focused in Scenario RT_III occupies an area of 4,500 ha. It consists of the construction of a 37 km long dike with a height of 6 m and the storage capacity of 2.7 million m^3. It is placed on the left bank of the Chi River from the vicinity of Yasothon weir to about 27 km upstream of Maha Chana Chai (E20A) gauging station;
- according to the aforesaid scenarios, some scenarios could be combined as it might yield the best results. In this case, Scenario RT_II+III represented a combination of Scenario RT_II and III.

It is important to note that the distance, which was used to specify the alignments of all retention basins from the reference points such as gauged sites and weirs, in this alternative referred to the main channel distance.

However, in some areas, the flow control is unable to deal with flood levels exceeding the levels of the bank like Retention basin RT_IV (Figure 6.20), that is why those specific sites are not considered. With a defined starting point, floodwater enters a retention basin through a trapezoidal-shaped flood diversion channel by overflow from a rising channel via a flap gate. While the flow re-entering a channel is regulated by adjustable gates, which operate based on pre-defined conditions.

Based on preliminary estimates, a usable retention space was estimated for each site (Table 6.13), while the calculated volume is based on various earth-fill dike heights and lengths with 1:1.5 side slopes and a top width of 4 m (Figure 6.21 and Table 6.14) (note: the figures provided in Table 6.13 and 6.14 are indicative and have some degree of uncertainty attached to them).

Table 6.13 Site descriptions of retention basin

Scenario	Retention area in ha	Dike height in m	Dike length in km	Retention volume in MCM
RT_I	5,500	9	43	500
RT_II	5,000	7	41	350
RT_III	4,500	6	37	270
RT_II+III	9,500	Depends on each site	78	620

Detailed flood modelling has been carried out in order to determine the effect that the retention basin flood mitigation performance will have on the flow characteristics of the watercourse. In this investigation, the impact that the upstream retention basin has on the flood flows as compared to the downstream ones. Of the scenarios considered above, Scenario RT_II, III, and II+III were found to provide little benefit in reducing flood damage to those properties previously affected by flooding. This may be explained by the fact that at the lower part of the Chi River Basin, retention basins will have only little potential beneficial impact on flood mitigation as they may hold floodwater that would normally be gone and release it afterward. However, the release of the floodwater may occur at the same time when the flood wave propagates along the Chi River reaches at the retention site. Consequently, the retention basin at such a location may eventually result in an increase in damaging floods. Due to their low versatility with negligible flood reduction capabilities, these scenarios are not recommended for inclusion in this study and would encourage upstream flow retardation rather than at the point of flooding. The greater benefits are likely to incur by an upstream retention basin (Scenario RT_I) as it can significantly decrease downstream flows and has a greater cumulative effect on associated downstream damages. In this instance, there is a 5,000 ha reduction in the approximate extent of the floods (Figures 6.22a and 6.22b).

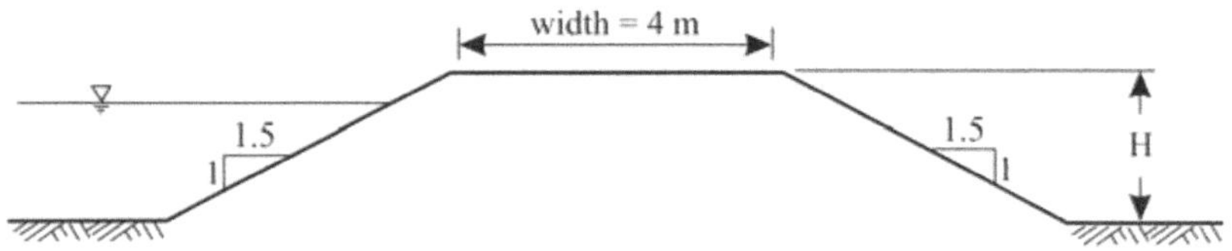

Figure 6.21 Cross section of dike retaining structure

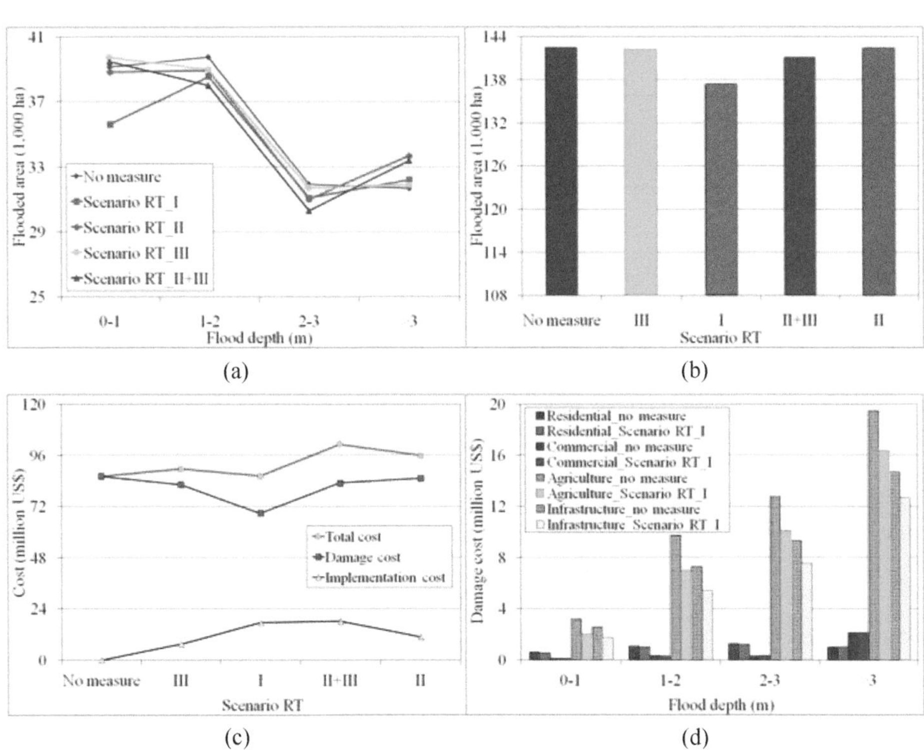

Figure 6.22 Four types of illustration of results from flood modelling under retention basin alternative: a) flooded area at each flood depth for all scenarios; b) flooded area under different scenarios; c) cost-benefit analysis of the scenarios; and d) damage cost per land use categories under the optimal scenario

Prior to making a final decision, an assessment of the impact of retention basins needs to be assigned. As indicated in Figure 6.22c, Scenario RT_I was found to be the best choice as it arrives at minimum total cost. Refer to Table 6.14, a cost breakdown is listed for each retention basin. According to this estimation, a majority of the cost is for the construction of the dike retaining structure, which is obtained by multiplying the quantity of earth-fill by the respective unit cost including its operation and maintenance, and estimated to be approximately US$ 17 million in the case of Scenario RT_I. If this is compared to the implementation cost, selected application would yield benefits up to US$ 17.3 million by reducing flood damage. From the view of land use, the comparison between subjects with and without the selected application scenario gives an illustration of the influence exerted by the upstream flood retention basin on the magnitude of flood extent. This may appear a little surprising as it does not have a substantial effect in reducing flood damage (Figure 6.22d). However, this alternative is worth implementing because the proposed construction site is located within the extent of a flood with a 1% annual probability of exceedance, whereby it is no longer needed to locate in the new tracts of land.

Most importantly, there will still be a concern in regard to the maintenance requirement such as trash and debris removal. If not, the retention basin will be ineffective as the control devices could become obstructed and will not function as desired.

Table 6.14 Financial evaluation of retention basin alternative for scenario screening (unit cost = 0.7 US$/m^3)

Scenario	Volume of earth-fill for dike retaining structure in million m^3	Cost in million US$			
		Implementation		Damage	Total
		Construction	O&M		
No measure	-	-	-	86	86
RT_I	6.8	5	12	69	86
RT_II	4.2	3	8	85	96
RT_III	2.9	2	5	82	90
RT_II+III	7.1	5	13	83	101

Note: O&M = Operation and maintenance

As this investigation comes to an end, the selected scenario is found not to provide a complete solution as the significant overland flooding problems are still apparent. That means some further endeavours are called for to determine a broader suite of flood mitigation efforts together with an insight into the hydraulic performance that most closely matches the actual site conditions.

6.3.5 Selection of potential intervention options

Aside from the above mentioned challenges, a significant possibility of success of flood fighting efforts will be seen when the chosen solutions are carefully selected, tailored, prioritized and assigned. This is because not every single mitigation action will fit every flooding problem and sometimes find it rather hard to choose the right one. In essence, the extent of flooding potential can be reduced through some flood mitigation measures, while other measures may increase the potential flood threat with detrimental impacts. Because of this, alternatives may be dropped from further consideration, if there are too many flaws. On the contrary, the alternatives, which are likely to satisfy most of the detailed objectives and higher levels measures of success, would be taken forward to the best suited mitigation options.

This stage involves the selection of potentially attractive options from the preliminary list examined in the identification stage as to compare their quantified flood reduction effects in light of the following main considerations:

- technical effectiveness, in view of effectiveness in reducing flood extent;
- feasibility, with a view to incorporating anticipated flood damage.

Regarding which of the options chosen might be more plausible than others in terms of technical performance, Figure 6.23 and Table 6.15 illustrate that flooded areas of the 1% annual probability of exceedance flood along the Chi River are consistently lower than the case of without measures concerned for all alternatives. (Note: the resulting plots in Figure 6.23 reflect the flooded areas and flood depths that would occur over the Chi River floodplain). These findings indicate that promising ways against flooding attacks are considered as acceptable.

If thought of in monetary terms, the estimated flood reduction benefits of various alternative features are presented in Table 6.15. From the calculation of the corresponding costs, alternatives seem to be viable and are of crucial importance to the success of tackling flood mitigation challenges in the Chi River Basin.

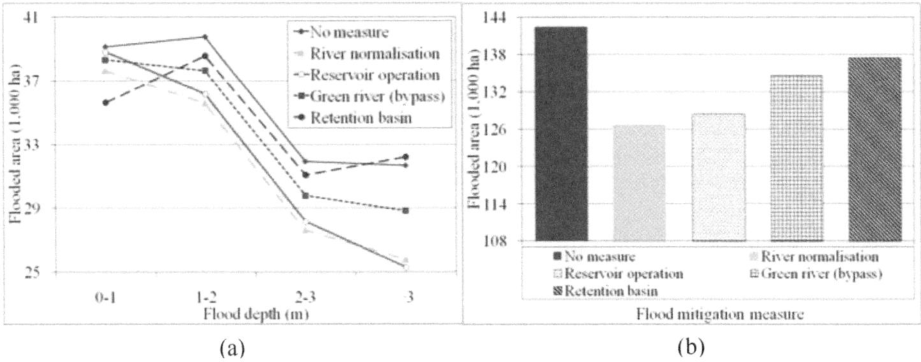

(a) (b)

Figure 6.23 Comparison of the inundated area with a 1% annual probability of exceedance from the four scenarios considered: a) each particular flood depth; and b) total flood extent

Table 6.15 Estimated flood reduction benefits of the identified flood mitigation alternatives

Alternative	Scenario	Implementation cost in million US$	Reduction	
			Damage in million US$	Flooded area in ha
No measure	No measure	-	-	-
River normalisation	RN_VI	21	22	16,000
Reservoir operation	RO_III	-	16	14,000
Green river (bypass)	GR_III+VI	9	11	8,000
Retention basin (upstream)	RT_I	17	17	5,000

Table 6.16 Identification of priority rank evaluated for individual alternatives

Evaluation criteria[1]	Alternative			
	River normalisation	Reservoir operation	Green river	Retention basin (upstream)
Feasibility[2]	**	****	***	*
Technical effectiveness[2]	****	***	**	*

Note: [1] = the evaluation criteria are not listed in order of importance
[2] = qualitative determination
* = very unfavourable
** = unfavourable
*** = favourable
**** = very favourable

The technical and financial insights seem well in line with each other. However, before a choice was made, the priority ranking system was assigned to those alternatives that meet the criteria. Table 6.16 shows the priority ranking of the four analysed measures based on the above discussions.

As illustrated in Table 6.16, it can be pointed out that the green river and retention basin alternatives are ranked as the third- and fourth-least desirable approach, respectively, as they are flawed in some way. Nevertheless, both are still the preferred remedial options, because obviously their adequacy and effectiveness outweigh their weaknesses. If this is decided, all the alternatives would be considered for the optimum combination of measures in the upcoming phases.

6.3.6 Optimal combination of generic intervention options

After the discussion on various measures for flood management and their effectiveness, now the question arises: what is the optimal solution of flood management? The answer is that none of them can be termed as the most advantageous, i.e. any alternative can be adopted in accordance with the circumstances. Therefore, effective responses may involve a suite or judicious combination of flood mitigation approaches rather than reliance on a stand-alone solution, because each of them will only transfer flood damage from one location to another and not solve the problem. Toward this end, it is necessary to bring together the most meritorious and desirable options of each alternative listed in Table 6.16, and sets out a range of combinations of measures with a high degree of interdependence on interaction processes between individual measures. However, the physical extent and consequences of floods, including the tendency towards flood dominance in the affected parts of the Chi River Basin, is not intended to be exhaustive owing to reasons given above. In this respect, the combination of complementary options that are being considered is discussed in subsequent paragraphs.

Mitigation measures matrix

The combination scenarios integrate the above mitigation options, since a single scenario sometimes induces effects opposite to others. The following matrix identifies various combinations of possible flood mitigation alternatives (Table 6.17). Eleven different combinations have been analysed for selection of the most promising combination, whereas the shaded areas indicate the consideration set for different scenarios, in an effort to prevent a repeat of the same scenarios.

Table 6.17 Matrix of potential measures for integrated model runs

Option	RN	RO	GR	RT	All
		Combination of the two measures			
RN	-	RN+RO	RN+GR	RN+RT	-
RO	RN+RO	-	RO+GR	RO+RT	-
GR	RN+GR	RO+GR	-	GR+RT	-
RT	RN+RT	RO+RT	GR+RT	-	-
		Combination of the three measures			
	RN+[RO+GR]	RO+[RN+GR]	GR+[RN+RO]	RT+[RN+RO]	-
	RN+[GR+RT]	RO+[GR+RT]	GR+[RO+RT]	RT+[RO+GR]	-
	RN+[RO+RT]	RO+[RN+RT]	GR+[RN+RT]	RT+[RN+GR]	-
		Combination of all four measures			
All	-	-	-	-	All

Note: RN = River normalisation (Scenario RN_VI)
RO = Reservoir operation (Scenario RO_III)
GR = Green river (Scenario GR_III+VI)
RT = Retention basin (Scenario RT_I)

Impact of scenarios on the threat of flooding

The alternatives that comprise the integrated model runs indicate that the manner in which combination scenarios are represented in the modelling approach has a significant impact on the flooding characteristics. To illustrate the magnitude of effects that could be achieved, quantitative comparisons have been made between scenarios. Table 6.18 and Figure 6.24a show that the combined impact of Scenario (RN_VI)+(RO_III)+(GR_III+VI)+(RT_I) provides the greatest reduction in the extent of 1% annual probability of exceedance flood of 41,100 ha. Since Scenario No measure represents no flood mitigation measures in place, it is employed as baseline condition.

Benefits of combination of flood mitigation efforts

The benefits are a combination of the effectiveness of the mitigation measures outlined above in reducing flood losses. Table 6.18 gives a summary of effects of combination scenarios in terms of the avoided damage in comparison to the baseline scenario. In each of the eleven combination scenarios, there is evidence that the estimated benefits of the various flood mitigation measures in terms of tangible savings are quite substantial.

Table 6.18 Avoided flood damage potential through the provision of various combinations

Scenario	Implementation cost in million US$	Flood extent in ha		Flood damage in million US$	
		Potential	Benefit	Potential	Benefit
No measure	-	143,000	-	86	-
RN+RO	21	113,000	29,000	65	21
RN+GR	30	123,000	20,000	60	26
RO+GR	9	123,000	20,000	77	9
RN+RT	38	122,000	21,000	69	17
RO+RT	17	122,000	20,000	69	17
GR+RT	26	130,000	13,000	70	16
RN+[RO+GR]	30	109,000	34,000	55	31
RT+[RN+RO]	38	108,000	35,000	58	28
RO+[GR+RT]	26	118,000	25,000	62	24
GR+[RN+RT]	47	118,000	25,000	60	26
RN+RO+GR+RT	47	101,000	41,000	38	48

The estimated benefits shown in Table 6.18 indicate that the potential flood damage avoided of Scenario RN+RO+GR+RT is the highest. Remaining combination scenarios are not recommended and identified as low priority due to limited flood mitigation benefit achieved. In light of its financial benefits, the expected value of flood damages is reduced by approximately US$ 48 million, which is more than it will cost for implementation (US$ 47 million). It means that this promising combination might be desirable in order to reduce overall risk. However, it needs to be noted that these related costs do not consider inflation rates, interest rates for bank loans or design lifetime of the mitigation options. Furthermore, the above analysis has also not taken into consideration indirect effects, losses, and their costs.

Selection and approval of recommended combination

In order to obtain the optimal combination, it is therefore necessary to find a cost-effective solution, i.e. total cost, for which the highest mitigation level and the highest return on invested dollars are found. The total costs are the sum of implementation costs and the expected value of the flood damage. The optimal performance of the preferred combination is found by minimising the total costs as shown in Figure 6.24b. As a result, the optimal (tailor-made) flood management package that is appropriate in mitigating the flood condition of Chi River Basin can be achieved.

Consideration of the eleven combination scenarios shows that some of them are inappropriate. The analysis of the above results reveals that the final selected combination of options is Scenario (RN_VI)+(RO_III)+(GR_III+VI)+(RT_I), which gives the lowest total cost could be considered more favourably as the most effective solution. A simple illustration in Figure 6.25 makes it clear that where all measures taken are preferably done.

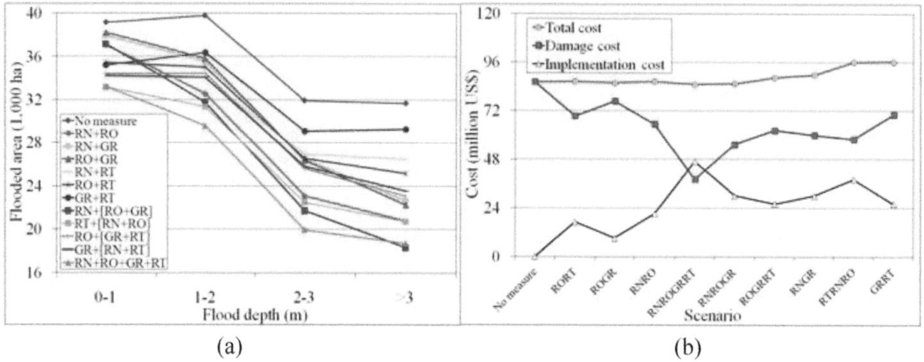

Figure 6.24 Two types of illustration of results from flood modelling under combination scenarios: a) comparison of the inundated area with a 1% annual probability of exceedance from eleven different combination scenarios at each particular flood depth; and b) cost-benefit analysis of the scenarios

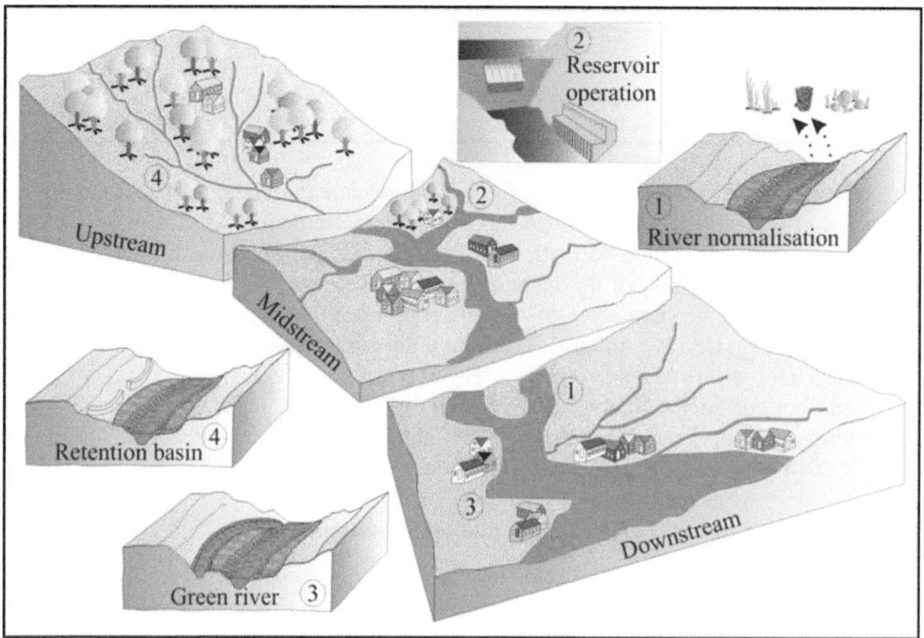

Figure 6.25 Recommended action items to tackle flood risk in the Chi River Basin

Taken as a whole, much of the merits of the right mix of approaches has now been realized. One piece of evidence that supports this remark is a significant reduction in flood damage for each flood affected land use element. A breakdown of the resultant reduction has been compiled, which relies upon their own unique damage characteristics, as detailed in Table 6.19. The damages shown in Table 6.19 represent the subsequent flood damages resulting from the specified flood event for each land use type and with respect to each flood depth. For instance, for the commercial land use type, some commercial areas are exposed to floods with flood depth between 1-2 m more than the commercial areas with 2-3 m of flood depth. It means that the affected areas with 1-2 m flood depth is larger than with 2-3 m flood depth, which results in difference in damage costs. This explanation also applies to the other land use types and different flood depths. A negative damage reduction value results from the damage transition from a deeper to a shallower flood depth, as well as from the situation that a decrease in flood damage in one particular land use type results in an increase in another. By some estimates, financial benefits from the selected application are found, resulting mainly from the reduction in the agricultural sector, contributing to approximately 54% of the gross value of damage reduction.

Table 6.19 Reduction in flood damage when the 1% annual probability of exceedance flood is reduced through optimum combination of measures

Flood depth in m	Land use type (2000 - 2002)	Damage in million US$		
		Without measure	With measures	Reduction
0 - 1	Residential	0.6	0.4	0.2
	Commercial	0.1	0.1	0.0
	Agriculture	3.2	2.1	1.1
	Infrastructure	2.6	1.7	0.9
1 - 2	Residential	1.1	0.7	0.4
	Commercial	0.4	0.5	-0.1
	Agriculture	9.8	5.2	4.5
	Infrastructure	7.3	4.1	3.2
2 - 3	Residential	1.3	0.6	0.7
	Commercial	0.3	0.2	0.1
	Agriculture	12.8	5.0	7.8
	Infrastructure	9.4	3.8	5.6
>3	Residential	1.0	0.3	0.7
	Commercial	2.1	1.0	1.2
	Agriculture	19.5	7.0	12.4
	Infrastructure	14.7	5.4	9.3
Total damage reduction				48.0

One more important point to add is that since the flood of extreme rarity (1% annual probability of exceedance) may take place at any time in a 100-year period and may turn violent without any prior warning, therefore, it is well worth elaborating on the different flood event occurrences in relation to the costs of implementation of optimal (tailor-made) flood management package. When the above mentioned flood event occurs early during the design period event, that means there may be a likelihood of another similar flood occurring in any given year within the remaining time frame. If this is the case, the more frequent flood events, the more benefits from the optimal solution to be gained, and vice-versa. In addition, if such flood does not arise, this possible solution is to be a useless overestimation and appears to be unnecessary.

Figure 6.26 demonstrates a technical solution once the most appropriate blend of the above alternatives are spread out in the Chi River Basin. In conclusion on this point, the range of viable solutions to the chronic flood problem is proposed to be put in place in a cohesive manner, and involve the following:

- *structural options*:
  - normalisation and improvement of the Chi River over a distance of approximately 331 km, extends from the Chi-Lam Pao confluence through the downstream end of the Chi River;
  - construction of two green river channels on the right bank floodplain of the Chi River with a total length of about 53 km. The proposed alignment of the first diversion structure is approximately 36 km long, starting near the Chi-Lam Pao confluence and ending at roughly 22 km downstream of Ban Tha Khrai gauged site (E18). The seventeen kilometres long channel (second diversion) would divert floodwater from a point approximately 27 km below streamflow gauging station Maha Chana Chai (E20A) through just 5 km upstream of Tat-Noi weir;
  - construction of the off-channel 5,500 ha upstream retention basin on the left bank floodplain of the Chi River. It is located between Ban Khai gauged site (E23) and about 22 km upstream of Ban Kaeng Ko (E21) gauged site.
- *non-structural option*: changes in operation rules of the two reservoirs, i.e. Ubol Ratana (UB) and Lam Pao (LP). During the flood season, the outflows release from UB and LP would be approximately 55% and 75% of their original daily outflows respectively, while the initial reservoir volumes are expected to be about 50% (UB) and 30% (LP) of their original initial volumes. (Note: other possibly effective non-structural measures, e.g. flood proofing, flood insurance, flood warning and preparedness, etc., were not investigated in this study).

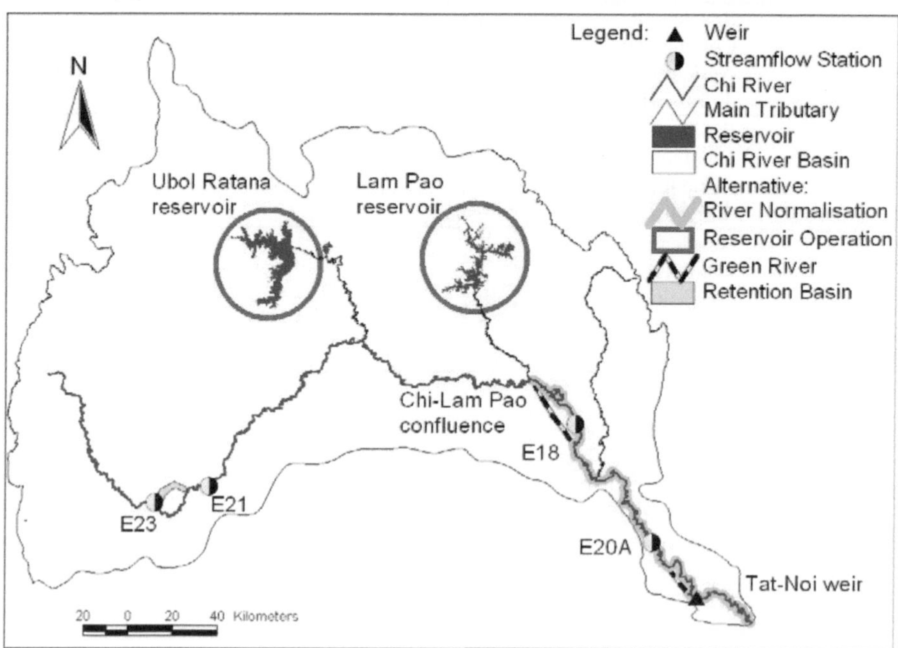

Figure 6.26 Schematic layout of the final selected combined scenario

By doing so, floodwater damage in the river basin can be eliminated or at least substantially reduced, and a better and safer living environment will be created. This statement indeed corresponds to the simulated results of flood inundation areas shown in Figure 6.27. The measures will be linked to the flood mitigation systems, which are already in place to complement and extend them.

Despite the large number of successful challenges in view of technical and financial perspectives from proactive flood mitigation efforts towards flood issues, the execution of proposed actions would also have various social and environmental impacts (e.g. Thompson and Sultana, 1996; Calder and Aylward, 2006), depending on case by case basis, i.e. the approach used and the current configuration of river channels. For instance, if there is a disturbance in river flow, it would cause sediment imbalance and possibly change the hydraulic processes of erosion and deposition. Consequently, sediment transport processes will continue to progress downstream which would likely result in negative effects on existing habitats, as well as substantial sediment accumulation and would require repeated sediment removal. Through a process involving the construction of impounding reservoirs, the consequent blocking of sediments will affect nutrient transport, and thereby nutrients may not reach downstream fertile farmlands. Meanwhile, disconnection of river and floodplain could also limit riparian vegetation development. In all this, both social and environmental impacts of floods and flood mitigation measures, as well as the way to incorporate them with technical and financial aspects would need to be accounted further.

Before closing this chapter, it can be seen in Figure 6.27 by the fact that the optimum combination of measures cannot prevent all floods, nor can it eliminate the possibility of loss if flooding occurs. For this reason, it is crucial to account and prepare for residual risks that follow the implementation of flood management measures. Therefore, further engineering solutions as well as non-structural measures would probably need to be included, for example the construction of a new dike apart from the existing dikes to further reduce the residual risk of flooding in the flood prone areas.

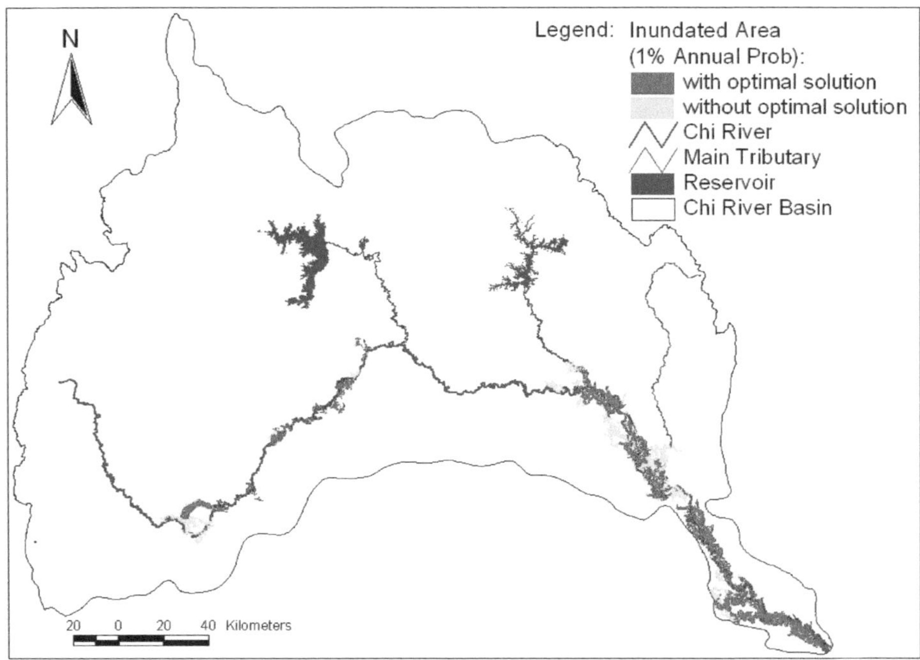

Figure 6.27 Comparison of inundated areas during a 1% annual probability of exceedance under the situation with and without optimum combination of measures concerned

7 Implications for future flood management actions in the Chi River Basin

7.1 Initiatives

The previous chapter showed the combined efforts to tackle effectively the flood threat with an annual probability of exceedance of 1%. However, the coincidence of the causes of floods in most cases lies beyond human control, and there is always the chance of a flood of greater severity occurring in the future. This concern will become worse if people's flood experiences and memories tend to fade quickly, by which it could compromise flood management issues and fail to address challenges/problems in a timely manner.

To answer the question, is there any actual solution to keep the area from future catastrophic floods once they would occur, the truth is that there is no such thing. However, there is still an unambiguous way to guide how to minimize possible negative effects of floods when they do arise. In this sense, the future implications of an optimal (tailor-made) flood management package, which was previously thought capable of adequate protection against risks to life and property, will inevitably need to be addressed in regard to future flood conditions under the influence of land use change. The firsthand efforts toward this point would probably help to concentrate on what is essential and without a lack of focus to react to future unexpected situations.

In line with the aforementioned recommendations, there is a need for decisive actions that are generally forward thinking, innovative, and action-oriented, focused, targeted and realistic. These include expeditious action together with supporting mechanisms, i.e. institutional arrangements and continuing competency assessment. The following is the progress made that will certainly be useful in inevitable future endeavours.

7.2 Anticipation of consequences of future flood risks and integrated flood management needs

Up until this point, the results appear to be persuasive enough to succeed in managing flood threats. However, it would be less likely to fully confide if flood anticipation, which explains the circumstances that have not yet come to light, is still missing. It is not a big deal whether flooding will happen or not, but the bigger issue is when it could happen. This concern is raised to confine acts of memories fade, distorted thinking and lack of readiness with the passage of time. Otherwise the level of public awareness on the impact of ordinary to catastrophic floods will become low and virtually unmanageable. In this situation, the emphasis would be on likelihood and potential impacts of future floods caused by projected increases in flood magnitude and changes in land use.

7.2.1 Future flood potential

While the flood tragedy still did not take place, why not prepare to draw a much more complete picture of potential future flood risk. This means that the magnitude of such damage might probably be inversely proportional to better anticipation. In particular, if more severe floods might occur, it is insufficient to consider only the 1% annual probability of exceedance flood. To address this issue, an insight into exceptionally large flood (e.g. 0.1% annual probability of exceedance flood) is required to achieve a higher 'optimal' safety level for dealing with crisis situations.

Illustrating possible worst-case conditions, the presentation below will provide a walk-through of taking appropriate actions rather than waiting for incidents to happen.

Since the type of flood that people usually hear about is the frequent flood, then when the extreme flood is considered, it may confuse their perception regardless either larger or smaller floods will likely occur. Therefore, the comparison of the inundation areas for flood with 0.1% and 1% annual probability of exceedance is presented in Figure 7.1, whereas no mitigation action would have been taken. It has to be stressed that 0.1% annual probability of exceedance flood is estimated with even larger uncertainty. In this study, it is assumed that the probability density function (PDF) for the rainfall can be extrapolated (stationarity assumption) to this extreme value. Note that no observations exist to support this 'rough' estimation. Furthermore, it is assumed that all flood mitigation measures will work properly even during this very extreme flood.

The slight differences become apparent when laying down both fringes of flooded areas, their envelopes seem largely similar. However, these differences will be put aside as their primary differences are the inundation depth. Plots demonstrating the outstanding differences in inundation at each particular flood depth are shown in Figure 7.2a in which it draws a clear distinction between dark blue and light green line segments. A large increase in the sum of all flood-affected areas is shown in Table 7.1.

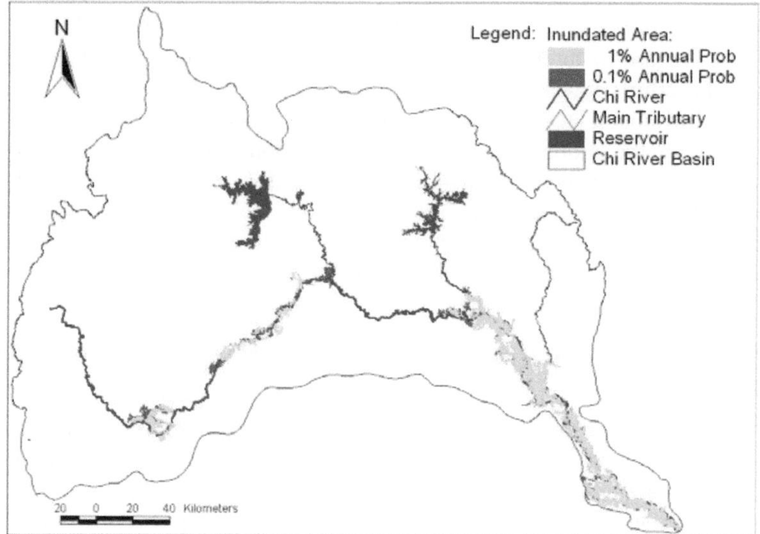

Figure 7.1 Comparison of inundated areas during 0.1% and 1% annual probability of exceedance floods

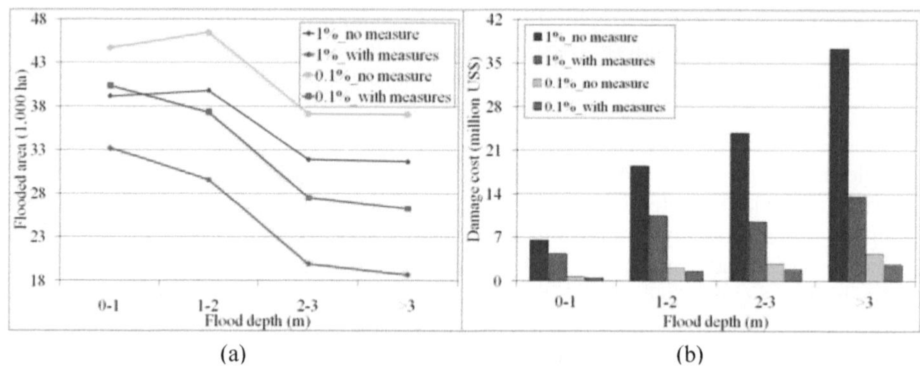

Figure 7.2 Two sets of results, during the 0.1% and 1% annual probability of exceedance flood events under present land use conditions: a) estimated flooded areas; and b) flood damage exposure

Table 7.1 Comparison of different flooded area for 1% and 0.1% annual probability of exceedance

Optimum combination measures	Land use	Flooded area in 1,000 ha		
		Annual probability of exceedance		Increase
		1%	0.1%	
Without	Present	143	165	15
With	Present	101	132	31

With these adverse consequences in mind, one would also have to consider the potential flood damage. This is illustrated in Figure 7.2b. This graphical representation visibly reveals that each flood depth has different expected flood damage cost for each of the aforesaid probabilities of exceedance levels. An explanation for this can be seen in Table 7.2. While the damage cost are computed for the 50 years period the damage cost for 0.1% annual probability of exceedance are lower than in case of 1% annual probability.

Table 7.2 Comparison of different flood damage for 1% and 0.1% annual probability of exceedance

Optimum combination measures	Land use	Damage in million US$	
		Annual probability of exceedance	
		1%	0.1%
Without	Present	86	10
With	Present	38	7

To make things clear, estimates with a difference in potential flood damage for different land use, which represents the number of sensitive and vulnerable assets, would necessarily have to be carried out. The subsequent flood damages resulting from the 0.1% annual probability of exceedance flood event are presented in Table 7.3. Significant increase in damage to agriculture appears to be larger than others, it comprises 52% of the total damage (note: in order to make the differentiation of flood rarity more easily understood, a current issue being discussed will need to refer back to Table 6.19 for a more detailed breakdown of flood damages caused by a 1% annual probability of exceedance flood).

Table 7.3 Estimated flood damages per land use type from the 0.1% annual probability of exceedance flood

Flood depth in m	Land use type (2000 - 2002)	Damage in million US$		
		Without measure	With measures	Reduction
0 - 1	Residential	0.1	0.1	0.0
	Commercial	0.0	0.0	0.0
	Agriculture	0.3	0.3	0.1
	Infrastructure	0.3	0.2	0.1
1 - 2	Residential	0.1	0.1	0.0
	Commercial	0.1	0.0	0.0
	Agriculture	1.1	0.8	0.3
	Infrastructure	0.8	0.6	0.2
2 - 3	Residential	0.2	0.1	0.1
	Commercial	0.0	0.1	0.0
	Agriculture	1.5	1.0	0.5
	Infrastructure	1.1	0.7	0.3
>3	Residential	0.1	0.1	0.1
	Commercial	0.2	0.1	0.1
	Agriculture	2.3	1.4	0.9
	Infrastructure	1.8	1.0	0.7
Total damage reduction				3.4

Particular attention is also given to the estimated monetary damage on a unit area basis since the expected potential damage alone does not clearly describe the severity of floods. Table 7.4 shows the flood damage costs per unit flooded area.

Table 7.4 Estimation of flood damage costs per unit area affected by floods

Depth in m	Damage cost per unit area affected in US$/ha			
	1%* no measure	1%* with measures	0.1%* no measure	0.1%* with measures
0 - 1	166	130	15	13
1 - 2	466	355	46	42
2 - 3	746	483	74	68
>3	1,180	734	121	101

Note: * Annual probability of exceedance
The above results are based on current land use conditions.

Estimated potential benefits through avoided flood damage

Based on the findings, increased pressure would rise to boost interest and attract attention in order to reduce future flood vulnerability in the Chi River Basin.

To illustrate this point, the expected value of avoided flood damage potential after finding an optimal (tailor-made) set of flood mitigation measures, is estimated and depicted in Figure 7.2 and Tables 7.1 to 7.4. In regard to this extraordinary flood, much of its potential damage could be avoided as shown in those tables. Behind this, the inclusion of implementation costs of the optimal solution would give an indication concerning this issue. Table 7.5 gives the cost breakdown of the tangible mitigation efforts.

By referring to the benefits of completed actions and their estimated costs, it is admitted that this set of efforts is justified even though the threat of flood becomes larger. However, despite significant potential savings have been identified, the potential for catastrophic damage still exists. People are particularly vulnerable as a result of over-reliance on overly optimistic mitigation measures under circumstances, which can actually create more harm than good. The reinforcement of a number of weak links in the flood alleviation schemes would need to be in harmony with the principle of 'learning to live with the floods'. Considerable efforts would have to be made by turning negative impacts of floods into positive aspects, and then the beneficiary aspects of floods will eventually bring new opportunities for livelihoods (Kuntiyawichai et al., 2011a).

Table 7.5 Summary of estimated implementation costs of each mitigation action

Mitigation component	Cost in million US$		
	Construction	O&M	Sum
River normalisation	6.0	15.0	21.0
Reservoir operation	-	-	-
Green river (bypass)	2.6	6.5	9.1
Retention basin	5.0	12.5	17.5
Total sum of completed mitigation action cost			47.6

Estimate of indirect adverse consequences of flood damage

Once the details have been listed, most of the images seem to be set out and captured. The earlier discussions posted that the direct flood damage is caused to objects directly by physical contact with the floodwater, while indirect flood damage, which occurs as a result of the aftermath of severe floods, was not explicitly described. Therefore, indirect flood damage estimation would need to be included in order to complete the full range of flood damages.

On one level, a direct consequence of flood disasters is easy to figure out in monetary value rather than an indirect one. To identify indirect effects, one must first look back to the estimated total direct damage (Tables 6.19 and 7.3) since the indirect monetary estimate can be estimated as a percentage of the direct flood damage. The following percentages were applied, as proposed by Kates (1965): 15% for residential land, 35% for commercial, 45% for industrial and 10% for agricultural. The results are displayed in Tables 7.6 and 7.7, which show the estimation of tangible direct and indirect potential damages for the 1% and 0.1% annual probability of exceedance flood, respectively. While not exhaustive, on the basis of the obtained results, the total tangible damage, which was calculated as the sum of monetary values of direct and indirect physical damages, is also given in Tables 7.6 and 7.7.

Table 7.6 Direct, indirect and total tangible impacts caused by the 1% annual probability of exceedance flood

Land use type (2000 - 2002)	Damage in million US$					
	Direct		Indirect		Tangible damage	
	no measure	with measures	no measure	with measures	no measure	with measures
Residential	4.0	2.0	0.6	0.3	4.6	2.3
Commercial	3.0	1.8	1.0	0.6	4.0	2.4
Industry*	-	-	-	-	-	-
Agriculture	45.2	19.3	4.5	1.9	49.8	21.3
Infrastructure	33.9	15.0	-	-	33.9**	15.0**
Total					92.3	41.0

Note: * No direct damage to industry is incurred under this flood and land use conditions
** These costs do not take into account its indirect effects since it is already included in the disruptions of economic and social activities of the other land use categories

Table 7.7 Direct, indirect and total tangible impacts caused by the 0.1% annual probability of exceedance flood

Land use type (2000 - 2002)	Damage in million US$					
	Direct		Indirect		Tangible damage	
	no measure	with measures	no measure	with measures	no measure	with measures
Residential	0.5	0.3	0.1	0.0	0.6	0.3
Commercial	0.4	0.3	0.1	0.1	0.5	0.4
Industry*	-	-	-	-	-	-
Agriculture	5.2	3.5	0.5	0.3	5.8	3.8
Infrastructure	4.0	2.6	-	-	4.0**	2.6**
Total					10.8	7.1

Note: * No direct damage to industry is incurred under this flood and land use conditions
** These costs do not take into account its indirect effects since it is already included in the disruptions of economic and social activities of the other land use categories

As an attempt to arrive at the complete damaging effect of floods on society, the resulting cost would have to include tangible and intangible damage assessment. However, an intangible damage assessment is widely recognized as very complicated and its inherent subjectivity makes it intrinsically hard to quantify. This arises with the statement that the analysis is often narrowed to the consideration of tangible monetary effects (Jonkman et al., 2004). Correspondingly, only tangible damage is considered sufficient to proceed to conduct a cost-effectiveness analysis, accordingly this study only acknowledges intangible damages but does not attempt to quantify them although their inclusion may improve the accuracy of the damage assessment.

7.2.2 Detecting and modelling flood consequences of land use dynamics

The particular issue here concerns the linkages between land use changes in the whole river basin, increased severity of floods and their interactions. It has received much attention since several crucial questions are of high scientific and societal interest (e.g. Ott and Uhlenbrook, 2004, Masih et al., 2011, etc.). As a precaution to broaden people's thought and action about the effects of anthropogenic alterations, a series of straightforward scenarios are discussed to illustrate its significant influence with the help from the calibrated models.

Future land use conditions in the Chi River Basin will not remain the same from what they are at present. Moreover, when they occur, it is very difficult to reverse to previous condition. These processes have the potential to influence on how floods are generated. Therefore, it will be necessary to project changes in land use. To illustrate this, the expected land use of 2057 has been used to identify what these changes tend to be, as well as how the Chi River Basin is likely to respond. This is based on the 'Northeastern Thailand Regional Plan'(Department of Public Works and Town & Country Planning, 2007).

Based on the detailed GIS data bases produced by the Department of Public Works and Town & Country Planning, Thailand, the potential effects of land use changes have been explored to measure the consequences of distributed changes in agricultural, forest, urbanisation and industrial land use. By referring to Figure 7.3, it visualises that a certain part of the agricultural land will be converted to other land use, which may gain quite significantly (Figure 7.4). Table 7.8 and Figure 7.5a show the area changes.

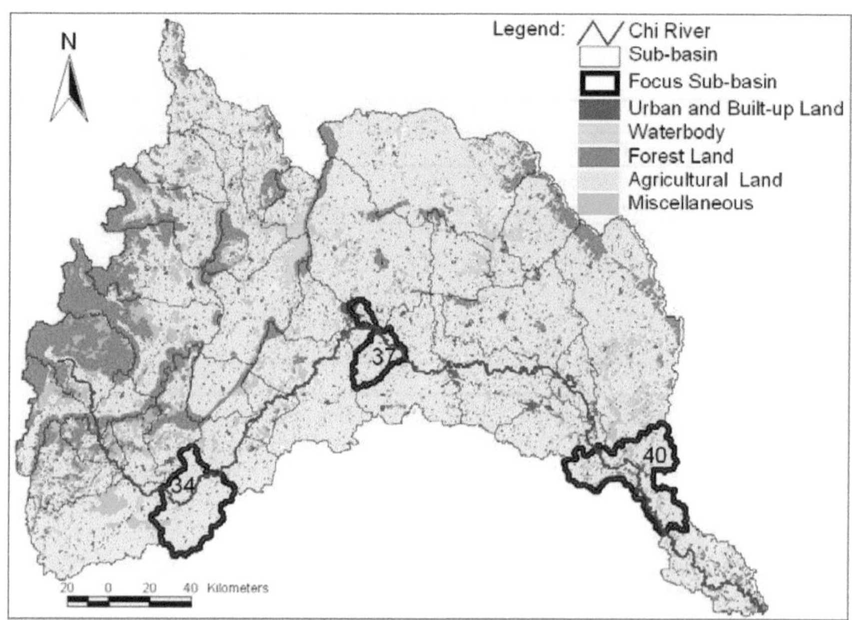

Figure 7.3 Baseline land use (2000 - 2002)

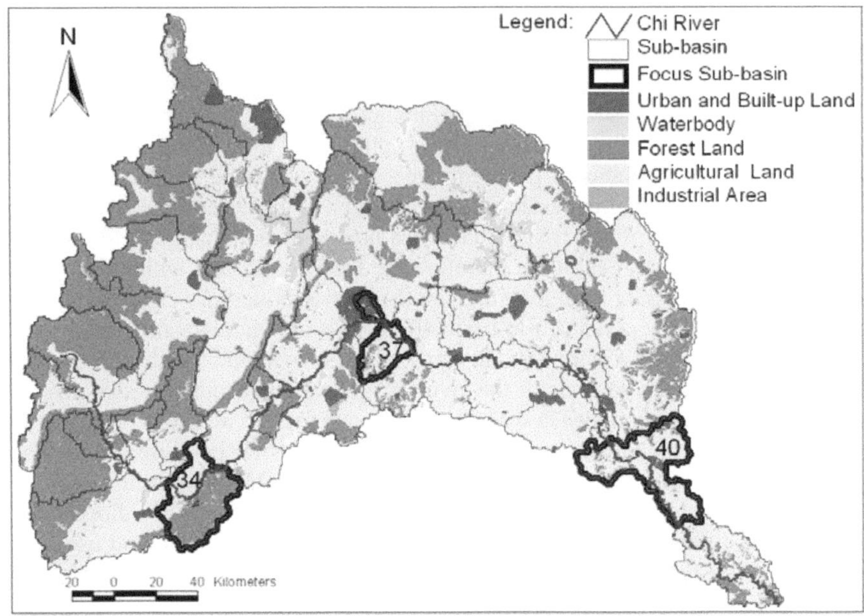

Figure 7.4 Changes in land use concerning planning scenario in 2057

Table 7.8 Percentage of land use area change for land use change impact modelling

Land use	Percentage of land use area		Percentage of change
	2000 - 2002	2057 (Planned)	
Agriculture	75.5	56.6	-18.9
Forest	17.9	37.2	+19.3
Urban	3.8	2.9	-0.9
Industry	0.1	0.6	+0.5
Waterbody	2.7	2.7	+0.0

As seen from the area changes, the preliminary figures reveal a reduction in urban areas, which is in the direction opposite to the normal trend. However, this might result from the tendencies towards higher density and more centralized urban planning. In addition, the increase in a wide range of industrial sites is another factor that can respond strongly to changing urban environment.

Pointing towards the potential impact of land use changes, it has been found that the consequent changes in runoff are generally small. In this sense, three selected sub-basins from different parts of the Chi River Basin are used to illustrate the land use change effects, i.e. sub-basin 34, 37 and 40 as the representations of upstream, midstream and downstream focus areas, respectively (Figure 7.3).

Careful thought on the upstream side (sub-basin 34), the consequences of the ongoing anthropogenic changes will have a positive impact since they appear to decrease the peak flow as seen in Figure 7.5b. This indicates that expanding areas of forest could result in alterations in flood regime so that the 1% annual probability of exceedance flood can be reduced. The evidence also corresponds with the statement of Bosch and Hewlett (1982) that the establishment of forest cover on sparsely vegetated land decreases water yield during floods (cf. also Ott and Uhlenbrook (2004), García-Ruiz et al. (2008), Saghafian et al. (2008), etc.).

In mid-river basin (sub-basin 37) and downstream regions (sub-basin 40), when the discharges are compared with each other, it seems that there is no significant difference (Figures 7.5c and 7.5d). Needless to say, the impact of these effects proved rather insensitive in each of those focus areas. Although one might argue that in sub-basin 37 the urban area is planned to be increased and centralized, then changes in flow regimes due to land use change should have emerged. However, since the forest area is also expected to be expanded, this results in compensation to increased flow caused by urban expansion. In the case of sub-basin 40, there is no significant change in future land use compared to the present condition, therefore the land use change within this sub-basin will not play an important role. Even though the upstream land use change may have an impact on runoff at the outlet of sub-basin 40, there is only a minor difference in runoff as can be clearly seen in Figure 7.5d. In this study, it is seen that the changes in land use do not have significant impacts at the river basin scale, but rather at sub-basin scale. This finding is in correspondence with the studies of Pfister et al. (2004) and Ward et al. (2008), which concluded that there is no evidence exists that land use change has had significant effects on peak flows at the river basin scale. Moreover, this result is also supported by Saghafian et al. (2008), who revealed that change in the flood peak and flood volume differs in different sub-basins because of non-uniform land use changes. Among these findings, it is necessary to pay more attention to those particular areas where they have a local impact, especially if significant developments are expected.

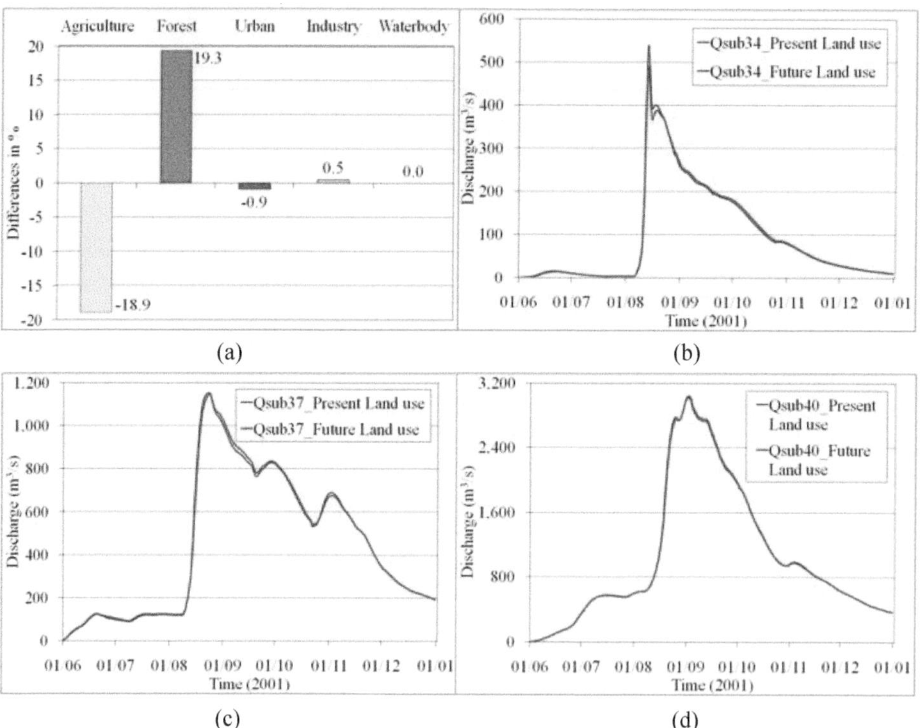

Figure 7.5 Results of the impact of projecting anthropogenic land use changes resulting from the 1% annual probability of exceedance design event: a) percent change in land use categories; b) alteration in upstream discharge (at outlet of sub-basin 34); c) difference in discharge at the mid-river basin (at outlet of sub-basin 37); and d) influence on downstream discharge (at outlet of sub-basin 40)

Considering the fact that determination of future flood damage is vital, it is considered essential to visualise the effects on potential flood damage. Although the results represent a rather small effect on flows, the flood damage would still be different because the land will be converted to other uses, as well as the more densely developed context. In this perspective, it is not only interesting to quantify damage in monetary terms, but also in the flood affected area.

Looking at the issue of flood inundation extent, taking as a baseline condition the current land use pattern, only a slight reduction in severely affected areas by a 1% annual probability of exceedance flood was found, i.e. from 143,000 ha to 142,000 ha (Figure 7.6a). This result can easily be misinterpreted and people might ignore its foreseeable consequences, therefore, a cognitive perspective towards efforts to understand the likelihood and severity of possible losses should not focus merely on the reduction in flooded areas but also its damage cost. In terms of monetary damages, since land exploitation increases as well as it changes, there is potentially more exposure to floods and thereby increasing the probable damage due to the most significant change in commercial property development sector at risk of flooding. To support this, the damage shown in Figure 7.6a then rises from US$ 86 million to as high as US$ 140 million. However, if this flood situation is well addressed, it will keep the effects of flood damage to a minimum. In dealing with a threat response, a thorough insight on how optimal flood damage mitigation efforts affect the possible future flood conditions is essential. Results of total expected damage presented in Figure 7.6a indicate that the decrease in potential damage with the application of optimal mitigation actions is about 40% (from US$ 140 million to US$ 85 million), and reduction in inundated area about 39,000 ha.

Not only that notable flood event described above, the flood with an annual probability of exceedance of 0.1% has also been considered. Similar to the results from 1% annual probability of exceedance flood, the future land use change only results in a small flooded area reduction from the case of existing land use conditions (Figure 7.6b). With regard to its damage, this would increase by 78% to US$ 18 million from US$ 10 million. Yet execution of an effective and thoughtful flood mitigation intervention ensures that the degree of flood severity is kept under control as seen by a decline of flood extent, which occurs over an area of 163,000 ha to 131,000 ha. Moreover, the results also emphasize the avoidance of damage up to US$ 6 million.

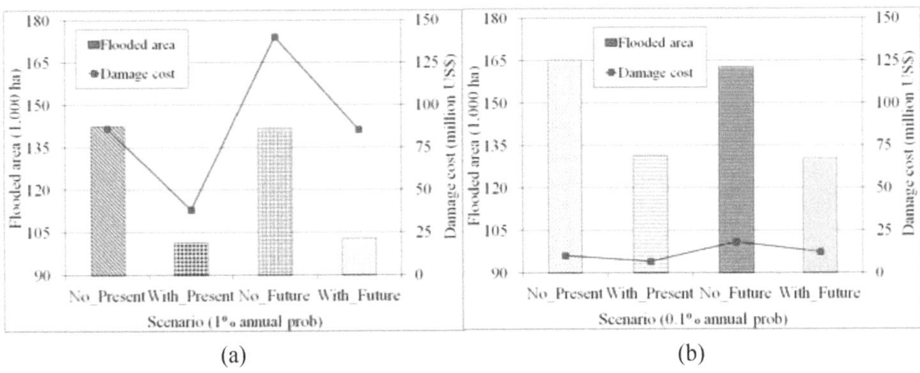

Figure 7.6 Potential flood damage resulting from land use change during flood events of specified annual probability of exceedance: a) 1%; and b) 0.1% (note that no/with mean no and with active mitigation efforts respectively, while present/future refer to land use conditions)

7.3 Integration of land use and flood management

7.3.1 General principles

Based on the results of this study, there is one best flood management solution, which is basically a feasible solution. Once completed, this finding will grasp a potentially problematic situation caused by floods and help to identify proper interventions in order to establish flood mitigation improvements. However, one thing that is nearly always true is that this finding is not always the end of seeking since reality seems always to deviate from the simulated conditions, which may be expected to occur.

Given the significance of how optimal integration of land use and flood management might proceed, it would be of particular importance to examine the extent to which possible actions would mitigate chronic flood problems.

Owing to the fact that inappropriate spatial planning can exacerbate the negative effects of extreme hydrological processes, flood losses appear to be increasing despite mitigation efforts, as more people and property are concentrated in locations at risk. As a result, flood likelihood and severity would appear to increase in the form of flood frequency, extent, and subsequent hazards.

To reduce risks to an acceptable standard, this section attempts to outline the principles of spatial planning options for adaptation to extreme flood events in order to incorporate this more effectively in flood loss reduction strategies for the Chi River Basin. Alternative spatial planning scenarios will take into consideration a series of management tasks to restrict flood prone areas to particular uses, specify where the uses may be located and establish minimum requirements for them, including the following objectives:

- to limit construction of structures and concentration of expensive goods and infrastructure on land subject to periodic inundation;
- to ensure that development maintains free passage and temporary storage of floodwaters in order to minimize flood damage;
- to ensure that the effect of inundation is not increased through development and will not cause significant rise in flood level or flow velocity;
- to minimize development and settlement in flood prone areas and prevent inappropriate development occurring in potentially hazardous areas;
- to conserve and maintain the productivity of prime crops.

The following steps would have to be applied to ensure that spatial planning offers an optimal solution, i.e. the development is appropriately designed and minimizes the need for redesigns:

- to identify areas, which would be affected by a flood event;
- to identify high hazard areas, which have the greatest risk and frequency of being affected by flooding.

Spatial planning should take into consideration the inputs from flood inundation, flood hazard and flood risk zone maps. Therefore, further steps will need to be explored on how the hydrological/hydraulic modelling outputs could be incorporated into spatial planning in order to develop the best possible (tailor-made) flood management package.

7.3.2 Non-structural flood management measures in floodplains and managing land use

The optimal package of flood management found so far cannot completely counteract flood risk in the Chi River Basin. The areas that are still known to be flood-liable along the Chi River as a result of the 1% annual probability of exceedance flood are mostly floodplains and low-lying areas starting from the Chi-Lam Pao confluence all the way to the Chi-Mun confluence. It is true that all affected areas are not exposed to the same flood depth. Therefore, it can be suggested that the focus of adverse impacts on flood behaviour and hazard and additional alternatives would have to be addressed simultaneously with regard to geographic location. In addition, flood management initiatives would also have to be people-centric, i.e. they need to react promptly and respond proactively to the needs of the communities, in particular the most vulnerable. The suggestion herein is a step forward, which includes enhanced efforts to cover the 'what if?' situations, and guide what people should do when their lives and livelihoods are disrupted by flooding (note: by assuming that a set of measures is already applied in the area). To be clear, one needs to outline how flood risk to existing and future developments can be managed.

Current floodplain developments

To consider the matter thoroughly, it is important to highlight that urban sprawl is fragmented and scattered across the Chi River Basin, while some of them are located in downstream flood prone areas (Figure 7.7). Once a flood occurs, the resulting potential flood damage can be tremendous. In response to these perceived risks, the relocation of residents could easily be seen as an alternative, however, it is difficult to put into practice as a result of the disagreement among local stakeholders and a high incentive/compensation is required in order to make it happen. If relocating people is not an option, informing the floodplain dwellers how to safeguard their lives and properties from flooding on their own initiative, is a matter which requires attention and is considered as an important action.

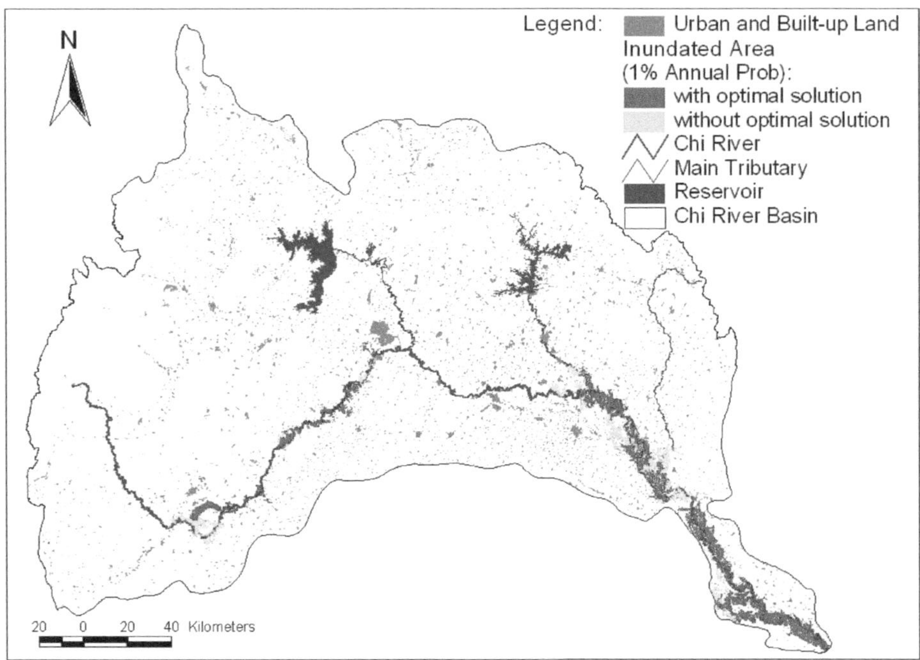

Figure 7.7 Present urban areas and simulated flood prone areas

Figure 7.8 shows the flood depth map, which classifies the inundation depth into 3 classes, i.e. 0.00 - 0.50 m, 0.50 - 2.00 m, and > 2.00 m, that leads to essential remarks on how to act when a flood comes. The depth classes are based on the Japanese Nationwide Standard (note: there is no such kind of standard in Thailand), which was reviewed by van Alphen et al. (2009). The actions in the following list are to be taken in relation to specific ranges of flood depth.

- flood depth between 0.00 - 0.50 m, flood proofing option can be applied in this specific situation, which includes the provision of temporary walls around individual buildings and raising the habitable floor levels above the flood level. In agricultural area, which is the most severely impacted sector, the modification of land management practices would be recommended (flood adaptation). That is to say, the farming should be restricted to the non-flood periods, only an appropriate type of crop that can withstand waterlogging and extended periods of inundation should be grown if the wet season farming is still needed due to economic reasons, and the construction of private dikes and stock refuges would also need to be included;
- flood depth between 0.50 - 2.00 m, the inhabitants would have to move to the first floor (the house has to be strong enough);
- flood depth more than 2.00 m, the evacuation of people off the floodplain in which the provision of timely and accurate information obtained through an effective flood warning system, has to be considered.

In any case, public parks or sport fields may also need to be used to increase the water retention capacity.

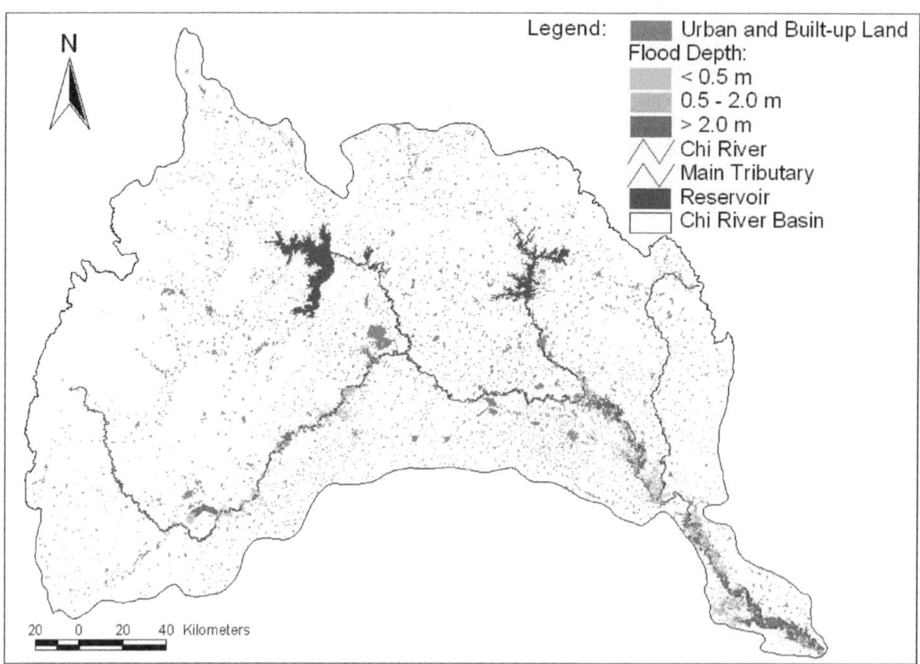

Figure 7.8 Flood depth map as a result of the optimal solution applied for present land use condition

Future floodplain developments

In the context of new developments, the Chi River Basin 2057 land use plan focuses on limiting the expansion of urban areas, which results in the reduction in urban flood prone areas (Figure 7.9). To adapt future land use planning to flood management while the optimal set of flood mitigation interventions is considered, it is proposed to allow for multifunctional land use by combining different functions that would deliver multiple benefits in a single area (Schultz, 2006a). In this perspective, designated flood storage areas can be used during flood-free months for recreational (non-residential) purposes, development of nature reserves, special types of agriculture, etc. In case future developments continue to take place in areas of high flood risk, such areas would need to ensure that the developments with residual flood risk are designed to be flood compatible and are limited to the purposes which are least threatening such as car parking, parks, outdoor sports facilities, etc. The intention of this guidance is also to ensure that new developments do not reduce the natural characteristics and capacity of floodplains. This also includes maximizing opportunities for the agricultural sector as well as ecological protection and restoration. Furthermore, it is important to note that climate change needs to be taken into consideration since its impacts can affect the results of flood characteristics, e.g. flood depth and flood extent, which may result in required adjustments of flood mitigation actions.

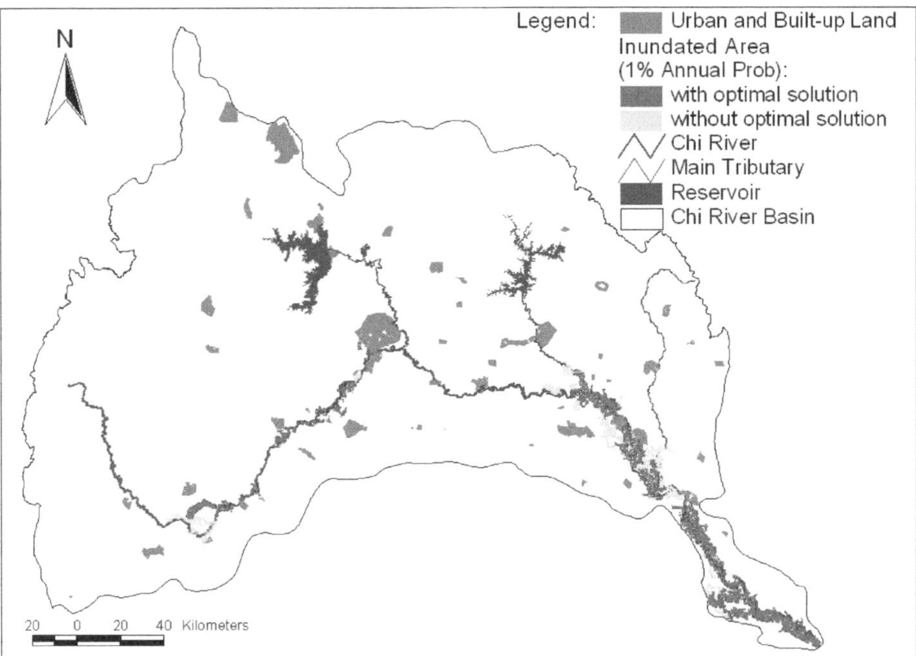

Figure 7.9 Future urban areas and simulated flood prone areas

It can be seen in Figure 7.10 that according to the future land use plan there are still some urban areas located in the high risk flood prone areas even though the optimal flood mitigation solution proposed is applied in this simulation. Therefore, these areas need to be reconsidered carefully regarding to flood depth classes mentioned in Section 7.3.1. Attention has to be paid to the buildings subject to inundation less than 2 m in which they should be built on elevated piers or earth mounds with the use of floodwater resistant materials. In the case of flood depth greater than 2 m, it is recommended to steer future development away from those flood zones.

To conclude, it is often thought that there are no risks after management measures are put in place. This is actually never true even though the approach to be adapted in implementing the flood management practice seems complete and the best solution is based on the known conditions and modelling capabilities. In addition, more actions remain to be done continuously as the flood management plan is not a static document and new needs will arise in the years to come. Moreover, any comprehensive plan would also require compromises in the implementation due to the involvement of stakeholders. Long time is needed for implementing changes, while people naturally are impatient to see things implemented. Another thing that makes the situation more complicated is the people's attitude since they always believe that the government is responsible for protecting them from flood damage and often decide to prepare themselves insufficiently. Therefore, it should also be considered that once a year a flood emergency exercise is done and that inhabitants know what to do when a flood is coming. Next to that, all the exercises and procedures have to be documented and put in the administration section of each municipality. The procedures have to be as simple as possible and easy to follow. If there is little cooperation with local communities and very difficult circumstances prevail, people will continue to suffer from the consequences of severe floods on a regular basis, and all endeavours of flood mitigation and management could result in money flushing down the drain.

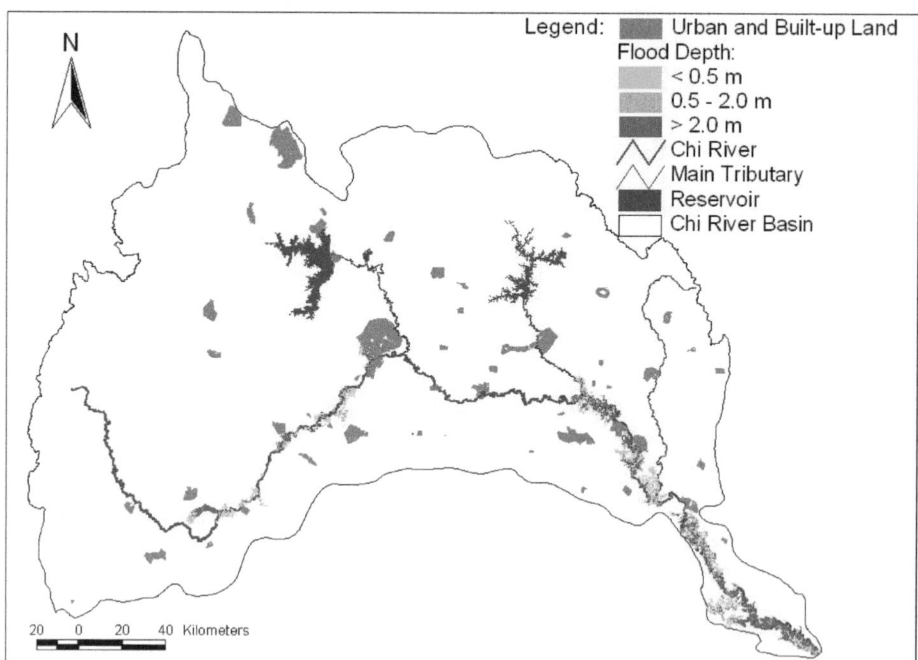

Figure 7.10 Flood depth map as a result of optimal solution applied for future land use condition

8 Evaluation

8.1 Before starting

On gathering together what have been raised from the threads of the foregoing chapters, those insights described will be linked to each other and pave the way for the evaluation of study findings. This chapter begins with a brief summary of key findings and conclusions for each of the objectives of this thesis. Then, the key challenges are addressed, and relevant recommendations together with directions for future research are also discussed. The chapter ends with a list of main lessons learned and closing remarks.

8.2 Wrap up

By adhering strictly and thoroughly to the identified success in all endeavours, it is time to bring a sense of closure. At this point, it needs to be noted that the findings were carried out under the main research objective, which was to develop a methodology to alleviate a recurrent flooding problem in the form of integrated modelling framework by taking into account the impact of the use of flood mitigation measures, land use change, and their subsequent effects on flooding behaviour. If the study achievement is further applied in practice, this would be a promising prospective and the identified flood problems would perhaps not be desperate.

8.2.1 Integrated modelling framework

The content presented in this thesis is considered to be a significant step forward in better understanding of coupling processes involved and modelling capabilities, giving a strong foundation for future research. In addition, it is also possible to extend the coupled components with other models for other applications.

A computational framework has been developed for representing the complex interactions of relevant hydrologic and hydraulic processes that led to extreme floods in the Chi River Basin. In this framework, efforts have been made towards an integrated interface system to couple the SWAT and 1D/2D SOBEK models through the Delft-FEWS platform. In more detail, the SWAT hydrological model was applied to investigate the effects of upstream river basin interventions and modelling the river basin response to changes by flood mitigation measures and land use. Further, the use of the 1D/2D SOBEK hydraulic model made it possible to simulate the complex flow patterns, and the maximum spatial extent of floods was mapped in which its presence would be highly informative about flood damages. The two composed models have a good predictive power as can be seen from the results that were closer to reality when calibrated and validated against observed discharges (SWAT model), and against measured water levels (1D/2D SOBEK model). The framework delivers good insights into likely conditions under specified probabilities of occurrence of any flood event, i.e. floods with a chance of occurrence of 0.1% and 1% per year. It captures a real detrimental effect to identify potential flood management opportunities. Application of this framework has been tested and applied. The results reveal that this integrated approach permits sensible inclusion and evaluation of different flood mitigation actions, which was previously not possible.

8.2.2 Identified flood prone areas

Based on historical observations and simulation results for previous flood events (year 2001), which were consistent and in reasonable agreement with each other, the most likely areas where devastating floods would take place are in the downstream part of the Chi River Basin (starting from the Chi-Lam Pao confluence through the Chi-Mun confluence). These consequences are in line with the absence of a downstream portion of the dike, as well as due to flat and low-lying topography, high rainfall intensities, insufficient upstream flood storage, restricted drainage efficiency of the Chi River, and improper land use in the floodplain area.

In case of floods with 0.1% and 1% annual probability of exceedance, the more destructive floods would occur in the upstream part, starting from Ban Khai gauged site (E23) through the Chi-Nam Phong confluence, and the Chi-Lam Pao confluence through the Chi-Mun confluence.

As such, determination of the likely flooding vulnerable areas is found very useful in identifying the shortcomings of the existing strategies for flood management in the Chi River Basin.

8.2.3 Coping with uncertainty

The recognition of uncertainty related to current and future flood management is important and needs to be appropriately considered. The following key points have arisen so far.

A little forethought is given to the effect of uncertainty on the estimation of infrequent floods, i.e. 0.1% and 1% annual probability of exceedance floods, which would need to be considered. In addition, it has to be realized that the integrated hydrologic and hydraulic modelling may be inherently accompanied by a high degree of uncertainty, i.e. all the errors of the processes involved in rainfall-runoff would probably be introduced in the overland flow process. It is not correct to claim that all sources of uncertainty such as model structure, input data, and parameter values are greatly minimized or eliminated through proper calibration. In addition, the results may also go wrong in an unpredictable way due to uncertainty associated with predicted land use changes. The uncertainty of flood damage assessment is also another source of error. In this sense, there may be inaccuracy since data on economic assets have become outdated, e.g. they are mainly based on data from 1989 and sites were located in Bangkok and surrounding areas, not in the Chi River Basin.

8.2.4 Applicability by end users

The state-of-the-art modelling framework for integrated flood simulation and management in the Chi River Basin seems to indicate the provision of a promising and viable alternative for end users. Since it integrates the strengths of two well-recognized modelling tools, i.e. SWAT and 1D/2D SOBEK, into a comprehensive package for all major aspects of flood modelling, it therefore enables the users to address the practical management of a river system and its riparian zone. If this integrated modelling framework is able to function well, the appraisal of flood mitigation options could be less time consuming and less modelling efforts which may result in the reduction of possible human errors. Thus, the full stand-alone package would increase the effectiveness of tailored interventions and action plans, as a result of a clear deliverable, which could be taken up quickly and effectively. This tool could also help to focus and concentrate on the efforts that can really have an impact in the final decision-making process. Furthermore, the knowledge required to build the models is not expected to be very high, which can plausibly convince interested non-specialists and provide them with useful and interesting insights.

8.2.5 Complete package for flood management

When a holistic flood management initiative has been launched, this is probably a way to unlock the flood potential inherent to river-floodplain systems and provides worthwhile benefits. It is not worth wasting time and effort on it, if the pressures on floodplains still continue to grow, together with the absence of mechanisms that control their development. Therefore, it is important to point out that development activities adjacent to riparian areas would have to be adjoined with appropriate interventions. Its invasion success is highly related to some of the points for consideration such as viable rather than functional extremity, active rather than passive response, and essentially far more low in cost with high returns.

Before thoughts turn towards actions, one would always need to keep clearly in mind that although the solutions for flood mitigation seem straightforward, they can likely result in the inevitable failures and are not absolute 100% safety guarantee if the improper management and maintenance are put in place as well as the lack of engineering and technical insight and understanding.

Prior to this, all flood mitigation efforts in the Chi River Basin, which are mostly not sufficiently concerned, entirely focused on structural-orientation with the view of stand-alone alternatives, which sometimes induce effects opposite to others, together with lack of integrated consideration across the river channel itself or its immediate vicinity. Hence, it is imperative that additional efforts need to be made as an integrated system through the implications of optimal feasible combination. At first, any one of the most common flood mitigation measures recommended today has been considered and its evaluation has been carried out using financial and technical efficiency criteria. These included structural (i.e. river normalisation, green river (bypass), and retention basin) and non-structural (e.g. reservoir operation) corrective alternatives (note: for each alternative, several possible scenarios were tested and evaluated to determine the most promising and efficient scenario). For those given alternatives, the damage potential as a consequence of flood depth with a 1% annual probability of exceedance has been considered as part of the appraisal. In order to meet this criterion, the beneficial reduction in expected damage must exceed the cost of implementation. Thus, the effectiveness in reducing flood inundation has also been taken into consideration on behalf of the technical performance criterion. Afterwards, the best case scenario for each alternative was identified in which its net benefits must be greater than the ones in other scenarios. The details can be shown in the following:

- normalisation and improvement of the Chi River, which is suggested to be carried out along its middle and lower reaches from the Chi-Lam Pao confluence till the end of the Chi River for approximately 331 km;
- changes in operation rules of the reservoirs. For Ubol Ratana reservoir, a change in the release quantity would have to be made to only 55% of its original daily outflow, together with a change to 50% of its initial volume at the beginning of the wet monsoon. In addition, a 75% outflow released and a 30% initial volume adjusted from its original initial volume would apply to the case of Lam Pao reservoir;
- construction of two green river channels on the right bank floodplain of the Chi River. The first one, approximately 36 km in length, starts from the Chi-Lam Pao confluence and ends at about 22 km downstream of Ban Tha Khrai gauged site

(E18). Another 17 km long diversion channel will take the floodwater from a point approximately 27 km downstream of Maha Chana Chai gauging station (E20A) through just 5 km upstream of Tat-Noi weir;
- construction of the off-channel 5,500 ha upstream retention basin, which is undertaken on the left bank floodplain of the Chi River between Ban Khai gauged site (E23) and nearly 22 km upstream of Ban Kaeng Ko (E21) gauged site.

Thereafter, a priority ranking system has been formulated and the results showed that all the alternatives can meet all screening criteria, which is indeed promising enough to be considered further. Presumably, since single alternatives sometimes induce effects opposite to others, therefore, all the alternatives have been brought to the forefront the need to consider an optimum combination of measures. The eleven combinations of scenarios were carried out in the context of combination of two, three and four alternatives. From exactly what they entailed and what were their benefits and costs, it was found that the optimum results can then be compiled by combining all four selected alternatives (based on the same aforementioned evaluation criteria). The results clearly demonstrated that such interventions are both financially and technically attractive since they can lead to significant expected reductions in impact severity of the 1% annual probability of exceedance flood, by more than their possible cost of implementation.

In addition, it is undoubtedly essential to increase the insight of flood behaviour related to future developments of the Chi River Basin. Merely pointing out the presence of arbitrary land use changes is not nearly as helpful as following the government land use planning (i.e. year 2057). In the plan, the amount of forest cover is predicted with a remarkable increase from 17.9% (2000 - 2002) to 37.2% (2057). This implies that the rise in forest cover will result in decline in agricultural areas (75.5% of the total river basin area) by up to 18.9% in 2057. The urban areas will also decrease 0.9% and this decrease will be replaced by the industry for approximately 0.5%. To demonstrate this point, the modelling results suggested that little or no significant potential impact of future land use change on river basin flood regime, but rather at sub-basin scale. It was apparently greater in some upstream sub-basins where additional forest land helps to buffer the effects of intensifying runoff, whereby there was no significant downstream hydrological impact. Certainly remarking on the massive flood damage, it will occur drastically due to higher density and more centralized urban growth. To contribute to future flood management, a simulation was conducted based jointly on future land use changes and optimal combination of flood mitigation measures. Their results indicated that optimal intervention could considerably reduce flood damage. Moreover, the 0.1% annual probability of exceedance flood event was also carried out afterwards, which found that its consequences corresponded closely to all the above findings.

By a flood damage determination for both considered design floods, it appears that the agricultural fields have the largest share in the total damage, as well as the largest area of inundation. Just as expected, infrastructure, residential and commercial areas are the second-, third- and fourth-highest total damage, respectively.

One more point, it would be better to ensure that future potential damage exposure will not bring about incorrect flood management decisions. To do so, the indirect costs were calculated and added to the direct costs, and these were just the tangible costs. A percentage of direct flood damage, i.e. 15% for residential land, 35% for commercial, 45% for industrial and 10% for agricultural, was accounted for as indirect costs. In this context, the results corresponded reasonably well to the overall substantive findings and practical suggestions.

After all, the anticipated impacts of growing development pressures, which are expected to outweigh the risks, would alter the exposure and vulnerability of the Chi River Basin, even though the optimum solution of complex flood problems already achieves. According to this issue, the explicit optimal integration of land use and flood management is increasingly recognized as an important item in shaping and controlling the flood damage potential within acceptable levels at all times, and needs to be carefully considered for further study.

8.3 Challenges ahead

Through the efforts of many, this study has met and successfully overcame many challenges. However, still some challenges lie ahead. Some of them can be listed as follows:

- the present thinking of holistic flood management that works with nature rather than against it, e.g. by restoring natural water retention capacities with the ability of wetlands and floodplains as natural sponges, as well as opposes the optimal solutions that rely primarily on just a single incentive mechanism, would have to be sought for;
- the prospect of further challenges also includes the presence of integrated efforts in relation to land use planning and management, which looks into the possibility of enclosing strengthening legislative and institutional arrangements;
- the most noteworthy achievement in flood management is more likely when cooperation from all involved parties in the Chi River Basin such as relevant authorities and stakeholders, non-profit organizations, and other agencies, is considered valuable or important. It is recognized that this may take time to ensure that it will be effectively acquired rather than inhibited;
- through the outstanding efforts in the successful completion of flood management solutions, it is thought that the level of mitigation action provided by them may not be considered adequate due to the potential increases in flood risk from climate change. Therefore, this is a crucial challenge and might also relate to others.

8.4 Recommendations

From the findings of this study, the following specific recommendations are presented to achieve truly integrated flood management.

- after completing the mitigation effort that has really made an acceptable difference, there is still residual risk remaining as the continuation of those endeavours induces tendencies and driving forces for further development, which does not follow the optimal use of space nor integrated spatial planning. Accordingly, reduction of exposure and vulnerability through spatial planning has to be strengthened. This needs to follow and definitely not denying the fact that some kinds of land use are more susceptible to floods than others. Moreover, the standard for spatial developments, which enables them to proceed with confidence and certainty, is also necessary to be considered;
- as disclosed therein, absolute security cannot be guaranteed while residual flood risks will still remain and everyone has to accept it. In order to handle the remaining as a result of the 1% annual probability of exceedance flood, it is recommended to build new dikes along the banks and the details are as follows:

  - between the Chi-Lam Pao and Chi-Nam Yang confluences, construction of the dike on the left bank as well as raising the elevation of the existing dike, the so-called 'Detch Chart', which also serves as the village road with 23.1 km in length, would be needed. On the right side, the existing dike, which is part of the Thung Saeng Badan scheme, would also have to be heightened;
  - the section from the Chi-Nam Yang confluence to downstream end of the Chi River, needs to be restricted by dikes on both sides;
- because one of the main of tasks of Ubol Ratana reservoir is hydropower generation, the reservoir needs to store as much water as possible in the reservoir. As a result, it might be difficult to release water before the rainy season in order to keep the space for the upcoming inflows for downstream flood mitigation. Even though it can be agreed to release it, there may be complaints from people living downstream of the reservoir since the huge amount of released water may destroy their properties, if reservoir water is released more than the conveyance capacity of the channel downstream of Ubol Ratana reservoir, which is estimated to be approximately 300 m^3/s. Hence, a public hearing, including certain clarifications that would increase understanding about the effectiveness of this particular management intervention in comparison to the effects of maintaining the amount of water stored in the reservoir, may be required;
- a thorough investigation of existing conditions of weirs along the Chi River, including its suitability for continued usage, is highly recommended to be implemented. This results from the fact that some weirs, e.g. Tat-Noi weir, have encountered problems during the flood periods in which the gate could not be opened and was still submerged to the depth of 50 cm;
- with regard to the issue associated with the modelled hydraulic characteristics of the area subject to flooding, it is indeed necessary to ensure the effective operation and maintenance of river and related flood management systems for their continued performance. Otherwise, there will likely be flooding at much lower flows in reality than in a modelled flood event, which still appears not to be at risk from flooding;
- within this, a breakthrough in other vegetative covers is also necessary in case food production goals would need to continue to achieve even during a flood. More emphasis can be given to such an extent that more flood-resistant/floating rice, which can tolerate submergence for a period of time without reducing yields, is widely grown in flood prone areas in the Chi River Basin. If less resistant crops are to be grown, crop cultivars that grow and ripen outside the flood season can be selected;
- one thoughtful recommendation would be appropriately presented here, which is closely related to political perspective. It could point to the fact that involved policy makers and persons sometimes are merely interested in the bigger and much costly flood management projects, even though there is extensive disagreement on this issue. This makes it necessary to change their attitudes towards putting in place in a balanced manner, and no longer attach to any fixed idea;
- the importance of local level, which is an actor that holds the real power to manage flood since they can understand very well the possible danger, cannot be denied. Therefore, consideration would have to be given to encourage their participation without hesitation, in accordance with clear institutional arrangements that can define responsibilities between different levels of government;

8.5 Future activities

Looking to the future, the following main points arise from this study, concerning elements that hold potential for subsequent developments.

One essential point would need to dedicate to the thought of storing floodwater for dry season use especially for irrigation, with the idea to make more double cropping possible. That is to say, there is a great potential to increase cropping intensity with the help of supplementary water supplies by construction of additional reservoirs, together with the augmentation of irrigation facilities and further fragmentation of river systems. Taken as a whole, additional simulations will then need to be made for this case in order to investigate the impact of multiple crops planted sequentially in combination with reservoir operation practices for flood mitigation.

Although a local action related to community involvement in flood management is no longer a fresh approach, upcoming phases of the development process can still take advantage of its positive effects. Therefore, scientific understanding and local/traditional knowledge such as local advice about safe locations and construction sites of flood mitigation measures would need to be taken together towards devising a flood management and mitigation programme.

The next matter to take into account is a more proactive and robust flood management response with consideration of the larger scale context. This idea will include integrated action in the neighbouring river basin, i.e. Mun River Basin. It means that a flood management initiative in the Chi River Basin should not increase flood risks in the aforementioned river basin, i.e. they need to be coordinated among each other.

In the task of continuation on strengthening synergy in integrated flood management actions, it is necessary to raise interest in the issue of precautionary through the process of flood forecasting and warning systems (note: extended from the modelling tool, which has been developed from this study). Prompt dissemination and regular surveillance, which enable the establishment of necessary counter measures, would be one of the core subjects and certainly encouraged.

The provision of an in-depth explanation in regard to the social concerns and environmental dimensions of flood vulnerability and exposure would also have to be included in order to ensure a truly integrated approach.

8.6 Main lessons learned and closing remarks

Drawing on the lessons learned from this study, it can be stated that the knowledge gained will bring a sound basis for a more comprehensive response to future flood incidents. The list of key messages below is a summary of the detailed lessons learned.

The integrated modelling framework provides a much more prudent method to identify where floods will occur. Based on this, flood behaviour can be understood ahead of time together with what the consequences will be, which will bring a detail of what the final solution best can be.

Part of the success of the simulations is due to the high numerical complexity, which is used to characterize the 1D/2D SOBEK hydraulic model. It helps to handle difficulties that may be encountered when simulating extreme flood situations. However, this complexity may require computational resources to complete the simulation processes.

True enough, floods are not merely the result of extreme natural phenomenon, but also the result of the interaction between the negligence of natural and social anthropogenic factors. Therefore, it is essential to consider the intervention increased perceptions and interest in flood risk as a whole and not just in parts of it.

There is no flood mitigation measure guaranteeing absolute safety, and sufficient flood mitigation safety cannot be reached in many vulnerable areas with the only help of structural measures due to technical, economic, and environmental constraints. The structural approaches are designed to protect only up to a certain level of flooding and can sometimes fail to reduce flooding or even increase economic losses from floods, as economic losses are merely postponed and continue to rise. In this sense, it may not always be a suitable solution and sometimes would be more effective if soft engineering techniques are applied or probably end up using both.

An integrated solution for flood management needs to be based on the ideas of solidarity in a holistic manner with involvement of representing the local conditions, together with the emphasis on land and water management problems as well as administrative responsibilities.

While floods can never be fully controlled, the beneficiary aspects of flooding are indeed appreciated as flood can bring new opportunities of livelihoods as well. Therefore, considerable efforts would have to be made by turning negative impacts of flood into positive aspects through the development of the notion of 'living with floods' or 'coping with floods'. To ensure a higher safety to floods, it is safer to 'give room to water' by restriction of human activities within the river basin, rather than relying on flood control infrastructure under all circumstances.

The issues, which have been addressed in this study, are the promise of flood management. However, serious flood-related problems may still exist due to the fact that too much emphasis on flood mitigation benefits sometimes led to negative impacts elsewhere. Therefore, the prescriptive answers presented should not be considered as a final solution, and it would be necessary to keep track of the whole system.

References

Amornsakchai, S., Annez, P., Vongvisessomjai, S., Choowaew, S., Thailand Development Research Institute, Kunurat, P., Nippanon, J., Schouten, R., Sripapatrprasite, P., Vaddhanaphuti, C., Vidthayanon, C., Wirojanagud, W., and Watana, E., 2000. Pak Mun dam, Mekong River Basin, Thailand, A WCD case study prepared as an input to the World Commission on Dams, Cape Town.

Ansusinha, K., 1989. Cost-benefit analysis of flood prevention programs for Bangkok and vicinity. Msc Thesis, Asian Institute of Technology, Bangkok, Thailand.

Arnold, J.G., Williams, J.R., Nicks, A.D., and Sammons, N.B., 1990. SWRRB: a basin scale simulation model for soil and water resources management. Texas A&M University Press, College Station, Texas, USA.

Arnold, J.G., Srinivasan, R., Muttiah, R.S., and Williams, J.R., 1998. Large area hydologic modeling and assessment part 1: model development. Journal of the American Water Resources Association, 34(1): 73 - 89.

Arnold, J.G., Allen, P.M., Volk, M., Williams, J.R., and Bosch, D.D., 2010. Assessment of different representations of spatial variability on SWAT model performance. Transactions of the American Society of Agricultural and Biological Engineers, 53(5): 1433 - 1443.

Asian Disaster Reduction Center, 1999. Thailand country report 1999, Kobe, Japan

Bank of Thailand, 2012. Thailand Floods 2011: impact and recovery from business survey, Economic Intelligence Team, Macroeconomic and Monetary Policy Department.

Bari, M.A. and Smettem, K.R.J., 2006. A conceptual model of daily water balance following partial clearing from forest to pasture. Hydrology and Earth System Sciences, 10(3): 321 - 337.

Barten, P.K. and Brooks, K.N., 1988. Modeling streamflow from headwater areas in the northern Lake States, In Modeling agricultural, forest, and rangeland hydrology. ASAE Publication 07-88, American Society of Agricultural Engineers, pp. 347 - 356.

Baulies, X. and Szejwach, G., 1997. LUCC data requirements workshop: survey of needs, gaps and priorities on data for land use/land-cover change research, Institut Cartogràfic de Catalunya, Barcelona, Spain

Beven, K.J., 2001. Rainfall-runoff modelling: the primer. John Wiley and Sons, Chichester, England.

Bicknell, B.R., Imhoff, J.C., Kittle, J.L., Donigian, A.S., and Johanson, R.C., 1997. Hydrological simulation program FORTRAN: user's manual for version 11. U.S. Environmental Protection Agency, Athens, Georgia, USA.

Boochabun, K., Vongtanaboon, S., Sukrarasmi, A., and Tangtham, N., 2007. Impact of land-use development on the water balance and flow regime of the Chi River Basin, Thailand. In: Sawada, H., Araki, M., Chappell, N.A., LaFrankie, J.V., and Shimizu, A. (Eds.), Forest environments in the Mekong River Basin. Springer Japan, pp. 24 - 35.

Borah, D.K. and Bera, M., 2003. Watershed-scale hydrologic and nonpoint-source pollution models: review of mathematical bases. Transactions of the American Society of Agricultural Engineers, 46(6): 1553 - 1566.

Borah, D.K. and Bera, M., 2004. Watershed-scale hydrologic and nonpoint-source pollution models: review of applications. Transactions of the American Society of Agricultural Engineers, 47(3): 789 - 803.

Borah, D.K., Bera, M., and Xia, R., 2004. Storm event flow and sediment simulations in agricultural watersheds using DWSM. Transactions of the American Society of Agricultural Engineers, 47(5): 1539 - 1559.

Bosch, J.M. and Hewlett, J.D., 1982. A review of catchment experiments to determine the effect of vegetation changes on water yield and evapotranspiration. Journal of Hydrology, 55(1-4): 3 - 23.

Brooks, K.N., Ffolliott, P.F., Gregersen, H.M., and Debano, L.F., 2003. Hydrology and the management of watersheds. Iowa State Press, A Blackwell Publishing Company, Ames, Iowa, USA.

Bruijnzeel, L.A., 1990. Hydrology of moist tropical forests and effects of conversion: a state of knowledge review. Humid tropics programme of IHP-UNESCO, Paris/Faculty of Earth Sciences, Free University Amsterdam.

Bultot, F., Dupriez, G.L., and Gellens, D., 1990. Simulation of land use changes and impacts on the water balance: a case study for Belgium. Journal of Hydrology, 114(3-4): 327 - 348.

Bureau of Reclamation, 1951. Irrigated land use, Part 2: Land classification, Bureau of Reclamation Manual vol. 5, U.S. Department of the Interior, U.S. Government Printing Office, Washington, DC, USA.

Calder, I.R. and Aylward, B., 2006. Forest and Floods. Water International, 31(1): 87-99.

Chow, V.T., 1959. Open channel hydraulics. McGraw-Hill, New York, USA.

Chow, V.T., Maidment, D.R., and Mays, L.W., 1988. Applied hydrology. McGraw-Hill, New York, USA.

Clawson, M. and Stewart, C.L., 1965. Land use information: a critical survey of US statistics including possibilities for greater uniformity. The Johns Hopkins press for resources for the future, Inc., Baltimore.

Collier, M.P., Webb, R.H., and Schmidt, J.C., 1996. Dams and rivers: primer on downstream effects of dams. U.S. Geological Survey Circular 1126, U.S. Government Printing Office, Washington, DC, USA.

Cornell University, 2003. SMDR The soil moisture distribution and routing model: documentation version 2.0. Soil and Water Laboratory, Biological and Environmental Engineering Department, Cornell University, Ithaca, New York, USA.

Costa-Cabral, M.C., Richey, J.E., Goteti, G., Lettenmaier, D.P., Feldkötter, C., and Snidvongs, A., 2008. Landscape structure and use, climate, and water movement in the Mekong River basin. Hydrological Processes, 22(12): 1731 - 1746.

Crews-Meyer, K.A., 2004. Agricultural landscape change and stability in northeast Thailand: historical patch-level analysis. Agriculture, Ecosystems and Environment, 101(2-3): 155 - 169.

Cunderlik, J.M., 2003. Hydrologic model selection for the CFCAS project: assessment of water resources risk and vulnerability to changing climatic conditions, Project report 1, University of Western Ontario, Canada.

De Laat, P.J.M., 1992. Workshop on hydrology lecture note, UNESCO-IHE, Delft, The Netherlands.

DeBarry, P.A., 2004. Watersheds: processes, assessment, and management. John Wiley and Sons, Hoboken, New Jersey, USA.

Delft Hydraulics, 2004. SOBEK: user's manual. Delft Hydraulics, Delft, the Netherlands.

Deltares, 2010. Delft-FEWS Documentation. Deltares, Delft, the Netherlands.

Department of Disaster Prevention and Mitigation, 2004. Handbook of the Ministry of Finance on funds allocation to emergency disaster relief assistance, Ministry of Interior, Bangkok, Thailand

Department of Disaster Prevention and Mitigation, 2005. Thailand's country report, Research and International Cooperation Bureau, Ministry of Interior, Bangkok, Thailand

Department of Public Works and Town & Country Planning, 2007. Northeastern Thailand regional plan, National and Regional Planning Bureau, Department of Public Works and Town & Country Planning, Bangkok, Thailand.

Department of Water Resources, 2003. Department of Water Resources and its role in Thailand's water management, Ministry of Natural Resources and Environment, Bangkok, Thailand.

Department of Water Resources, 2005a. First memory: 2003 Department of Water Resources, Ministry of Natural Resources and Environment, Bangkok, Thailand.

Department of Water Resources, 2005b. Main report, feasibility report of the integrated plan for water resources management in Chi River Basin, Ministry of Natural Resources and Environment, Bangkok, Thailand (in Thai).

Department of Water Resources, 2006. Main report, feasibility report of the integrated plan for water resources management in Chi River Basin, Department of Water Resources, Bangkok, Thailand (in Thai)

Department of Water Resources, 2007. Strategic plan for water resources management in 25 river basins, Executive summary, Ministry of Natural Resources and Environment, Bangkok, Thailand (in Thai)

DHI Water and Environment, 2007a. MIKE FLOOD 1D-2D Modelling: user manual. DHI Water and Environment, 108 pp.

DHI Water and Environment, 2007b. MIKE 11 a modelling system for rivers and channels: reference manual. DHI Water and Environment, 516 pp.

Dhondia, J.F. and Stelling, G.S., 2002. Application of one dimensional-two dimensional integrated hydraulic model for flood simulation and damage assessment. In: Falconer, R.A., Lin, B., Harris, E.L., and Wilson, C.A.M.E. (Eds.), Proceedings of the 5th International Conference in Hydroinformatics, Cardiff, UK, pp. 265 - 276.

Douben, K.J., 2006. Flood management lecture note. UNESCO-IHE Institute for Water Education, Delft, the Netherlands.

Douben, K.J. and Ratnayake, R.M.W., 2006. Characteristic data on river floods and flooding: facts and figures. In: van Alphen, J., van Beek, E., and Taal, M. (Eds.), Proceedings of the 3rd International Symposium on Flood Defence. Taylor & Francis/Balkema, Nijmegen, the Netherlands.

du Plessis, L.A. and Viljoen, M.F., 1999. Determining the benefits of flood mitigation measures in the lower Orange River: a GIS application. Water SA, 25(2): 205 - 213.

Duan, Q., Sorooshian, S., and Gupta, V., 1992. Effective and efficient global optimization for conceptual rainfall-runoff models. Water Resources Research, 28(4): 1015 - 1031.

Economic and Social Commission for Asia and the Pacific, 1991. Manual and guidelines for comprehensive flood loss prevention and management, United Nations Development Programme, United Nations, New York, USA.

Economic and Social Commission for Asia and the Pacific, 1999. Regional cooperation in the twenty-first century on flood control and management in Asia and the Pacific, United Nations, New York, USA.

Economic and Social Commission for Asia and the Pacific, 2007. Guidelines on integration of water-related disaster preparedness and mitigation into socio-economic development process, United Nations Development Programme, United Nations, New York, USA.

El-Nasr, A.A., Arnold, J.G., Feyen, J., and Berlamont, J., 2005. Modelling the hydrology of a catchment using a distributed and a semi-distributed model. Hydrological Processes, 19(3): 573 - 587.

Electricity Generating Authority of Thailand, 1997. Flood management and safety review of the Ubol Ratana dam, Executive summary, Bangkok, Thailand.

Eswaran, H., Beinroth, F.H., and Virmani, S.M., 2000. Resource management domains: a biophysical unit for assessing and monitoring land quality. Agriculture, Ecosystems and Environment, 81(2): 155 - 162.

Floch, P., Molle, F., and Loiskandl, W., 2007. Marshalling water resources: a chronology of irrigation development in the Chi-Mun River Basin, Northeast Thailand, M-POWER Working Paper MP-2007-02, Unit of Social and Environmental Research, Chiang Mai University, Thailand.

Food and Agriculture Organization and United Nations Environment Programme, 1998. Terminology for integrated resources planning and management. Soil Resources, Management and Conservation Service, FAO Land and Water Development Division.

Ford, D.T. and Hamilton, D., 1996. Computer models for water-excess management. In: Mays, L.W. (Ed.), Water resources handbook. McGraw-Hill, New York, USA

Frank, E., Ostan, A., Coccato, M., and Stelling, G.S., 2001. Use of an integrated one dimensional-two dimensional hydraulic modelling approach for flood hazard and risk mapping. In: Falconer, R.A., and Blain, W.R. (Eds.), River Basin Management. WIT Press, Southampton., UK, pp. 99 - 108.

Fukui, H. and Hoshikawa, K., 2003. Earthen bund irrigation in Northeast Thailand, In: Proceedings of the 1st International Conference on Hydrology and Water Resources in Asia Pacific Region, Kyoto, Japan, pp. 179 - 184.

García-Ruiz, J.M., Regüés, D., Alvera, B., Lana-Renault, N., Serrano-Muela, P., Nadal-Romero, E., Navas, A., Latron, J., Martí-Bono, C., and Arnáez, J., 2008. Flood generation and sediment transport in experimental catchments affected by land use changes in the central Pyrenees. Journal of Hydrology, 356(1-2): 245 - 260.

Gardiner, J.L., 1994. Sustainable development for river catchments. Water and Environment Journal, 8(3): 308 - 319.

Gardiner, J.L., 1998. Developments in floodplain risk management: decision-making in England and Wales, In: Floodplain risk management, Proceedings of International Workshop on Floodplain Management. Balkema Publishers, Rotterdam, the Netherlands, Hiroshima, Japan, pp. 291 - 306

Gassman, P.W., Reyes, M.R., Green, C.H., and Arnold, J.G., 2007. The Soil Water and Assessment Tool: historical development, applications, and future research directions. Transactions of the American Society of Agricultural and Biological Engineers, 50(4): 1211 - 1250.

Gassman, P.W., Arnold, J.G., Srinivasan, R., and Reyes, M., 2010. The worldwide use of the SWAT model: technological drivers, networking impacts, and simulation trends, In: Proceedings of the 21st Century Watershed Technology: improving water quality and environment Universidad EARTH, Guacimo, Costa Rica.

Goldstein, R.A., Mankin, J.B., and Luxmoore, R.J., 1974. Documentation of PROSPER: a model of atmosphere-soil-plant-water flow, Report EDFB-IBP-73-9, Oak Ridge National Laboratory, Oak Ridge, Tennessee, USA.

Guertin, D.P., Barten, P.K., and Brooks, K.N., 1987. The peatland hydrologic impact model: development and testing. Nordic Hydrology, 18: 79 - 100.

Holtan, H.N., Stiltner, G.J., Henson, W.H., and Lopez, N.C., 1975. USDAHL-74 model of watershed hydrology, Technical Bulletin No. 1518, Agricultural Research Service, United States Department of Agriculture

Hori, H., 2000. The Mekong: environment and development. United Nations University Press, USA.

House of Representatives' Ad Hoc Committee on Solutions to Water Resources, 2008. The study report on Thailand's water resources management framework, The House of Representatives, Royal Thai Government, Thailand.

Huising, E.J., 1993. Land use zones and land use patterns in the Atlantic zone of Costa Rica: a pattern recognition approach to land use inventory at the sub-regional scale, using remote sensing and GIS, applying an object-oriented and data-driven strategy. PhD Thesis, Wageningen University, Wageningen, the Netherlands.

Hydrocomp Inc., 1976. Hydrocomp simulation programming: operations manual. Palo Alto, California, USA.

Intarachai, T., 2003. Toward a knowledge-based economy: Northeastern Thailand. In: Makishima, M., and Suksiriserekul, S. (Ed.), Human resource development toward a knowledge-based economy: the case of Thailand. Institute of Developing Economies, Japan External Trade Organization, Chiba, Japan.

International Commission on Irrigation and Drainage, 1999. Manual on non-structural approaches to flood management, New Delhi, India

Jewitt, G., 2005. Chapter 186: Water and Forests. In: Anderson, M.G. (Ed.), Encyclopedia of Hydrological Sciences. John Wiley and Sons, West Sussex, UK, pp. 1 - 15.

Jonkman, S.N., Brinkhuis-Jak, M., and Kok, M., 2004. Cost benefit analysis and flood damage mitigation in the Netherlands. HERON, 49(1): 95 - 111.

Kangsheng, W., 2005. Long-term freshwater input and sediment load from three tributaries to Lake Pontchartrain, Louisiana, School of Renewable Natural Resources, Louisiana State University, Baton Rouge, , Louisiana, USA.

Kates, R.W., 1965. Industrial flood losses: damage estimation in the Lehigh Valley, Department of Geography Research Paper No. 98, University of Chicago, The University of Chicago Press, Chicago.

Kazi Emran Bashar, 2005. Floodplain modelling in Bangladesh by SOBEK 1D2D coupling system. MSc Thesis, UNESCO-IHE, Delft, the Netherlands.

Keys, C., 2004. Warning people about coming floods: recent developments and some barriers to improved performance, In: Proceedings of the 44th Annual Conference of the Floodplain Management Authorities of NSW, Coffs Harbour, New South Wales, Australia.

Krutilla, J.V., 1966. An economic approach to coping with flood damage. Water Resources Research, 2(2): 183 - 190.

Kuntiyawichai, K., Schultz, B., Uhlenbrook, S., and Suryadi, F.X., 2007. Application of the SWAT model in the Chi River Basin, Thailand, for different land use scenarios, A peer-reviewed poster presentation at the 4th International SWAT Conference, Delft, The Netherlands.

Kuntiyawichai, K., Schultz, B., Uhlenbrook, S., and Suryadi, F.X., 2008a. Delineation of flood hazards and risk mapping in the Chi River Basin, Thailand, In: Proceedings of the 10th International Drainage Workshop of ICID Working Group on Drainage, Helsinki/Tallinn, Finland/Estonia, pp. 298 - 313.

Kuntiyawichai, K., Schultz, B., Uhlenbrook, S., Suryadi, F.X., and van Griensven, A., 2008b. Importance of land use management on the flood management in the Chi River Basin, Thailand, In: Proceedings of the 4th International Symposium on Flood Defence, Toronto, Canada.

Kuntiyawichai, K., Schultz, B., Uhlenbrook, S., and Suryadi, F.X., 2009. Integrated hydrologic, hydraulic, and damage analysis for selection of optimal flood management measures, In: Proceedings of the 7th Annual Mekong Flood Forum, Bangkok, Thailand, pp. 199 - 209.

Kuntiyawichai, K., Schultz, B., Uhlenbrook, S., Suryadi, F.X., and Corzo, G.A., 2010a. Comprehensive flood mitigation and management in the Chi River Basin, Thailand, In: Proceedings of the 7th International Symposium on Lowland Technology, Saga, Japan, pp. 314 - 322.

Kuntiyawichai, K., Schultz, B., Uhlenbrook, S., Suryadi, F.X., and Werner, M., 2010b. Appropriate flood mitigation framework through structural and non-structural measures for the Chi River Basin, Thailand, In: Proceedings of the 8th Annual Mekong Flood Forum, Vientiane, Lao PDR.

Kuntiyawichai, K., Schultz, B., Uhlenbrook, S., Suryadi, F.X., and Corzo, G.A., 2011a. Comprehensive flood mitigation and management in the Chi River Basin, Thailand. Lowland Technology International, 13(1): 10 - 18.

Kuntiyawichai, K., Schultz, B., Uhlenbrook, S., Suryadi, F.X., and van Griensven, A., 2011b. Comparison of flood management options for the Yang River Basin, Thailand. Irrigation and Drainage, 60(4): 526 - 543.

Lakanavichian, S., 2001. Forest policy and history in Thailand: Working Paper No. 9, Research Centre on Forest and People in Thailand, Thailand.

Leaf, C.F. and Brink, G.E., 1973. Hydrologic simulation model of Colorado subalpine forest, USDA Forest Service Research Paper RM-107, Rocky Mountain Forest and Range Experiment Station, Fort Collins, Colorado, USA.

Lebel, L., Manuta, J., and Garden, P., 2011. Institutional traps and vulnerability to changes in climate and flood regimes in Thailand. Regional Environmental Change, 11(1): 45 - 58.

Lekuthai, A. and Vongvisessomjai, S., 2001. Intangible flood damage quantification. Water Resources Management, 15(5): 343 - 362.

Li, W., 2006. Dynamic analysis of water resources distribution in the Heihe River. MSc Thesis, UNESCO-IHE, Delft, the Netherlands.

Liu, Z. and Todini, E., 2002. Towards a comprehensive physically-based rainfall-runoff model. Hydrology and Earth System Sciences, 6(5): 859 - 881.

López-Moreno, J.I., Beguería, S., and García-Ruiz, J.M., 2006. Trends in high flows in the central Spanish Pyrenees: response to climatic factors or to land-use change? Hydrological Sciences Journal, 51(6): 1039 - 1050.

López-Moreno, J.I., Vicente-Serrano, S.M., Moran-Tejeda, E., Zabalza, J., Lorenzo-Lacruz, J., and García-Ruiz, J.M., 2011. Impact of climate evolution and land use changes on water yield in the Ebro basin. Hydrology and Earth System Sciences, 15(1): 311 - 322.

Lorsirirat, K., 2007. Effect of forest cover change on sedimentation in Lam Phra Phloeng reservoir, Northeastern Thailand. In: Sawada, H., Araki, M., Chappell, N.A., LaFrankie, J.V., and Shimizu, A. (Ed.), Forest Environments in the Mekong River Basin. Springer Japan, pp. 168 - 178.

Luo, Y. and Sophocleous, M., 2011. Two-way coupling of unsaturated-saturated flow by integrating the SWAT and MODFLOW models with application in an irrigation district in arid region of West China. Journal of Arid Land, 3(3): 164 - 173.

Manuta, J., Khrutmuang, S., Huaisai, D., and Lebel, L., 2006. Institutionalized incapacities and practice in flood disaster management in Thailand. Science and Culture, 72: 10 - 22.

Masih, I., Maskey, S., Uhlenbrook, S., and Smakhtin, V., 2011. Impact of upstream changes in rain-fed agriculture on downstream flow in a semi-arid basin. Agricultural Water Management, 100(1): 36 - 45.

Mays, L.W., 2005. Water resources engineering. John Wiley and Sons, New Jersey, USA.

McCuen, R.H., 2003. Modeling hydrologic change: statistical methods. CRC Press, Boca Raton, Florida.

Mekong River Commission, 2005. Overview of the hydrology of the Mekong Basin, Mekong River Commission, Vientiane, Lao PDR.

Ministry of Natural Resources and Environment, 2008. Integrated water resources management in Thailand, Bangkok, Thailand.

Moriasi, D.N., Arnold, J.G., Van Liew, M.W., Bingner, R.L., Harmel, R.D., and Veith, T.L., 2007. Model evaluation guidelines for systematic quantification of accuracy in watershed simulations. Transactions of the American Society of Agricultural and Biological Engineers, 50(3): 885 - 900.

Munich Reinsurance Company, 1997. Flooding and insurance, Munich, Germany.

Munich Reinsurance Company, 1998. World map of natural hazards, Munich, Germany.

Myers, M.F., 1997. Trends in floods, Workshop on the social and economic impacts of weather, Boulder, Colorado, USA.

Nash, J.E. and Sutcliffe, J.V., 1970. River flow forecasting through conceptual models part I: a discussion of principles. Journal of Hydrology, 10(3): 282 - 290.

Neitsch, S.L., Arnold, J.G., Kiniry, J.R., and Williams, J.R., 2005. Soil and Water Assessment Tool: theoretical documentation version 2005. Soil and Water Research Laboratory and Blackland Research Center, Temple, Texas, USA.

Nesbitt, H., Johnston, R., and Solieng, M., 2004. Mekong River water: will river flows meet future agriculture needs in the lower Mekong Basin? In: Seng, V., Craswell, E., Fukai, S. and Fischer, K. (Ed.), Water in agriculture, ACIAR Proceedings, pp. 86 - 104.

Office of Natural Resources and Environmental Policy and Planning, 2005. Thailand State of Environment Report 2005, Ministry of Natural Resources and Environment, Bangkok, Thailand.

Office of the National Water Resources Committee, 2000. National water vision: a case study of Thailand.

Ohara, N., Kavvas, M.L., Kure, S., Chen, Z.Q., Jang, S., and Tan, E., 2011. Physically based estimation of maximum precipitation over American River Watershed, California. Journal of Hydrologic Engineering, 16(4): 351 - 361.

Ott, B. and Uhlenbrook, S., 2004. Quantifying the impact of land-use changes at the event and seasonal time scale using a process-oriented catchment model. Hydrology and Earth System Sciences, 8(1): 62 - 78.

Pattanee, S., 2006. Challenges in managing the Chao Phraya's water, In: Proceedings of the 9th International Riversymposium, Managing rivers with climate change and expanding populations, Brisbane, Australia.

Pattanee, S., 2008. Implementation of IWRM in Thailand, Development Plan (BDP) Stakeholder Consultation Forum, Vientiane, Lao PDR.

Pawattana, C., Tripathi, N.K., and Weesakul, S., 2007. Floodwater retention planning using GIS and hydrodynamic model: a case study for the Chi River Basin, Thailand, In: Proceedings of the 6th International Conference on Environmental Informatics, Bangkok, Thailand, pp. 548 - 556.

Pfister, L., Kwadijk, J., Musy, A., Bronstert, A., and Hoffmann, L., 2004. Climate change, land use change and runoff prediction in the Rhine-Meuse basins. River Research and Applications, 20(3): 229 - 241.

Pikounis, M., Varanou, E., Baltas, E., Dassaklis, A., and Mimikou, M., 2003. Application of the SWAT model in the Pinios River Basin under different land use scenarios, In: Proceedings of the 8th International Conference on Environmental Science and Technology, Lemnos, Greece.

Pollution Control Department, 2004. Thailand state of environment: the decade of 1990s, Ministry of Natural Resources and Environment, Bangkok, Thailand.

Quiroga, C.A., Singh, V.P., and Lam, N., 1996. Land use hydrology. In: Singh, V.P., and Fiorentino, M. (Eds.), Geographical Information Systems in Hydrology. Kluwer Academic Publishers, Dordrecht, pp. 389 - 414.

Refsgaard, J.C. and Storm, B., 1995. MIKE SHE. In: Singh, V.P. (Ed.), Computer models of watershed hydrology. Water Resources Publications, pp. 809 - 846.

Refsgaard, J.C. and Knudsen, J., 1996. Operational validation and intercomparison of different types of hydrological models. Water Resources Research, 32(7): 2189 - 2202.

Regional Centre for Geo - Informatics and Space Technology, 2001. E-Sarn flood information system 2001, Khon Kaen University, Khon Kaen, Thailand

Royal Irrigation Department, 1988. Chi Basin water use study, Ministry of Agriculture and Cooperatives, Bangkok, Thailand.

Royal Irrigation Department, 1999. Chi River Basin blueprint, Ministry of Agriculture and Cooperatives, Bangkok, Thailand.

Royal Irrigation Department, 2006. Strategic report of water resources development in the Chi River Basin, Ministry of Agriculture and Cooperatives, Bangkok, Thailand (in Thai).

Royal Irrigation Department, 2008. Water management in Chi and Mun River Basin, Ministry of Agriculture and Cooperatives, Bangkok, Thailand (in Thai).

Royal Irrigation Department and Department of Water Resources, 2008. Improvement of irrigation efficiency on paddy fields in the lower Mekong basin project (IIEPF), Final report, Mekong River Commission.

Royal Irrigation Department, 2009. Lam Pao reservoir storage efficiency improvement project, Ministry of Agriculture and Cooperatives, Bangkok, Thailand (in Thai).

Saghafian, B., Farazjoo, H., Bozorgy, B., and Yazdandoost, F., 2008. Flood intensification due to changes in land use. Water Resources Management, 22(8): 1051 - 1067.

Sahasakmontri, K., 1989. Estimation of flood damage functions for Bangkok and vicinity. MSc Thesis, Asian Institute of Technology, Bangkok, Thailand.

Saleh, A. and Du, B., 2004. Evaluation of SWAT and HSPF within BASINS program for the upper North Bosque River watershed in central Texas. Transactions of the American Society of Agricultural Engineers, 47(4): 1039 - 1049.

Santhi, C., Arnold, J.G., Williams, J.R., Dugas, W.A., Srinivasan, R., and Hauck, L.M., 2001. Validation of the SWAT model on a large river basin with point and nonpoint sources. Journal of the American Water Resources Association, 37(5): 1169 - 1188.

Schultz, B., 2001. Irrigation, drainage and flood protection in a rapidly changing world. Irrigation and Drainage, 50(4): 261 - 277.

Schultz, B., 2006a. Opportunities and threats for lowland development. Concepts for water management, flood protection and multifunctional land-use, In: Proceedings of the 9th Inter-Regional Conference on Environment-Water, EnviroWater 2006, Concepts for Water Management and Multifunctional Land-Uses in Lowlands, Delft, the Netherlands

Schultz, B., 2006b. Flood management under rapid urbanisation and industrialisation in flood-prone areas: a need for serious consideration. Irrigation and Drainage, 55(S1): S3 - S8.

Schumann, A.H. and Schultz, G.A., 2000. Detection of land cover change tendencies and their effect on water management. In: Schultz, G.A., and Engman, E.T. (Eds.), Remote Sensing in Hydrology and Water Management. Springer-Verlag, Berlin, Germany.

Seibert, J., 1997. Estimation of parameter uncertainty in the HBV model. Nordic Hydrology, 28(4/5): 247 - 262.

Seibert, J., 2005. HBV light version 2: user's manual. Department of Physical Geography and Quaternary Geology, Stockholm University, Sweden.

Sethaputra, S., Thanopanuwat, S., Kumpa, L., and Pattanee, S., 2001. Thailand's water vision: a case study. In: Ti, L.H., and Facon, T. (Eds.), From vision to action: a synthesis of experiences in Southeast Asia, The FAO-ESCAP Pilot Project on National Water Visions, Bangkok, Thailand, pp. 71 - 98.

Shaw, E.M., 1994. Hydrology in practice, 3rd edition. Chapman & Hall, London, UK.

Singh, J., Knapp, H.V., Arnold, J.G., and Demissie, M., 2005. Hydrological modeling of the Iroquois river watershed using HSPF and SWAT. Journal of the American Water Resources Association, 41(2): 343 - 360.

Singh, S.K., 2004. Simplified use of gamma-distribution/Nash model for runoff modelling. Journal of Hydrologic Engineering, 9(3): 240 - 243.

Singh, V.P., 1989. Hydrologic systems: watershed modeling, Vol 2. Prentice Hall, New Jersey, USA, 320 pp.

Smith, P.J., 2005. Probabilistic flood forecasting using a distributed rainfall-runoff model, PhD Thesis, Graduate School of Engineering, Kyoto University, Kyoto, Japan.

Srinivasan, M.S., Gérard-Marchant, P., Veith, T.L., Gburek, W.J., and Steenhuis, T.S., 2005. Watershed scale modeling of critical source areas of runoff generation and phosphorus transport. Journal of the American Water Resources Association, 41(2): 361 - 377.

Srinivasan, R., Zhang, X., and Arnold, J.G., 2010. SWAT ungauged: hydrological budget and crop yield predictions in the upper Mississippi River Basin. Transactions of the American Society of Agricultural and Biological Engineers, 53(5): 1533 - 1546.

Stephens, K.A., van Duin, B., van der Gulik, T., Deong, Y., Maclean, L., and Bozic, L., 2005. The water balance model for Canada: improving the urban landscape through inter-provincial partnerships, In: Proceedings of the International Water Association Conference. Watershed and Basin Management-Sustainable Urban Drainage, Calgary, Alberta, Canada.

Suiadee, W., 2002. Organizational change for participatory irrigation management, Asian Productivity Organization, Tokyo, Japan.

Sun, G., Zuo, C., Liu, S., Liu, M., McNulty, S.G., and Vose, J.M., 2008. Watershed evapotranspiration increased due to changes in vegetation composition and structure under a subtropical climate. Journal of the American Water Resources Association, 44(5): 1164 - 1175.

Swedish Meteorological and Hydrological Institute, 2006. The HBV model. http://www.smhi.se/sgn0106/if/hydrologi/hbv.htm.

Thai National Mekong Committee, 2004. Sub-area study and analysis, 5T Sub-area, Basin Development Plan Unit, Department of Water Resources, Bangkok, Thailand.

The MathWorks Inc., 2008. Matlab 7.7.0. Natick, Massachusetts, USA.

Thompson, P.M. and Sultana, P., 1996. Distributional and Social Impacts of Flood Control in Bangladesh. The Geographical Journal, 162(1): 1-13.

Tramblay, Y., Bouvier, C., Ayral, P.A., and Marchandise, A., 2011. Impact of rainfall spatial distribution on rainfall-runoff modelling efficiency and initial soil moisture conditions estimation. Natural Hazards and Earth System Scinces, 11(1): 157 - 170.

Twine, T.E., Kucharik, C.J., and Foley, J.A., 2004. Effects of land cover change on the energy and water balance of the Mississippi River Basin. Journal of Hydrometeorology, 5(4): 640 - 655.

Uhlenbrook, S. and Leibundgut, C., 2002. Process-oriented catchment modelling and multiple-response validation. Hydrological Processes, 16(2): 423 - 440.

Uhlenbrook, S., Roser, S., and Tilch, N., 2004. Hydrological process representation at the meso-scale: the potential of a distributed, conceptual catchment model. Journal of Hydrology, 291(3-4): 278 - 296.

Uhlenbrook, S., 2006. Catchment hydrology—a science in which all processes are preferential. Hydrological Processes, 20(16): 3581 - 3585.

Uhlenbrook, S., 2007. Biofuel and water cycle dynamics: what are the related challenges for hydrological processes research? Hydrological Processes, 21(26): 3647 - 3650.

US Army Corps of Engineers, 2000. Hydrologic Modeling System (HEC-HMS): technical reference manual. US Army Corps of Engineers, Hydrologic Engineering Center, 149 pp.

van Alphen, J., Martini, F., Loat, R., Slomp, R., and Passchier, R., 2009. Flood risk mapping in Europe, experiences and best practices. Journal of Flood Risk Management, 2(4): 285 - 292.

van der Ent, R.J., Savenije, H.H.G., Schaefli, B., and Steele-Dunne, S.C., 2010. Origin and fate of atmospheric moisture over continents. Water Resources Research, 46(9): W09525.

van Duivendijk, J., 2005. Manual on planning of structural approaches to flood management. International Commission on Irrigation and Drainage, New Delhi, India.

van Griensven, A., Meixner, T., Grunwald, S., Bishop, T., Diluzio, M., and Srinivasan, R., 2006. A global sensitivity analysis tool for the parameters of multi-variable catchment models. Journal of Hydrology, 324(1-4): 10 - 23.

van Griensven, A. and Meixner, T., 2007. A global and efficient multi-objective auto-calibration and uncertainty estimation method for water quality catchment models. Journal of Hydroinformatics, 9(4): 277 - 291.

Van Liew, M.W., Arnold, J.G., and Garbrecht, J.D., 2003. Hydrologic simulation on agricultural watersheds: choosing between two models. Transactions of the American Society of Agricultural Engineers, 46(6): 1539 - 1551.

Verwey, A., 2001. Latest developments in floodplain modelling-1D/2D integration, In: Proceedings of the 6th Conference on Hydraulics in Civil Engineering, Hobart, Australia.

Viney, N.R., 2003. Modelling surface water in the Ord River Irrigation Area (ORIA), Technical report 39/03, CSIRO Land and Water, Perth, Australia.

Wagner, S., Kunstmann, H., and Bardossy, A., 2006. Model based distributed water balance monitoring of the White Volta catchment in West Africa through coupled meteorological-hydrological simulations. Advances in Geosciences, 9: 39 - 44.

Walsh, S.J., Evans, T.P., Welsh, W.F., Entwisle, B., and Rindfuss, R.R., 1999. Scale-dependent relationships between population and environment in northeastern Thailand. Photogrammetric Engineering and Remote Sensing, 65(1): 97 - 105.

Walsh, S.J., Crews-Meyer, K.A., Crawford, T.W., Welsh, W.F., Entwisle, B., Rindfuss, R.R., Millington, A.C., and Osborne, P.E., 2001. Patterns of change in land use, land cover, and plant biomass: separating intra- and inter-annual signals in monsoon-driven Northeast Thailand. GIS and Remote Sensing applications in biogeography and ecology, 626: 91 - 108.

Ward, P.J., Renssen, H., Aerts, J.C.J.H., van Balen, R.T., and Vandenberghe, J., 2008. Strong increases in flood frequency and discharge of the River Meuse over the late Holocene: impacts of long-term anthropogenic land use change and climate variability. Hydrology and Earth System Sciences, 12(1): 159 - 175.

Wenninger, J., Uhlenbrook, S., Tilch, N., and Leibundgut, C., 2004. Experimental evidence of fast groundwater responses in a hillslope/floodplain area in the Black Forest Mountains, Germany. Hydrological Processes, 18(17): 3305 - 3322.

Werner, M., van Dijk, M., and Schellekens, J., 2004. DELFT-FEWS: an open shell flood forecasting system. In: Liong, S., Phoon, K., and Babovic, V. (Eds.), In: Proceedings of the 6th International Conference on Hydroinformatics. World Scientific Publishing Company, Singapore, pp. 1205 - 1212

Werner, M. and Heynert, K., 2006. Open model integration: a review of practical examples in operational flood forecasting. In: Gourbesville, P., Cunge, J., Guinot, V., and Liong, S. (Eds.), In: Proceedings of the 7th International Conference on Hydroinformatics, Nice, France.

Werner, M. and Whitfield, D., 2007. On model integration in operational flood forecasting. Hydrological Processes, 21(11): 1519 - 1521.

Woodall, D.L. and Lund, J.R., 2009. Dutch Flood Policy Innovations for California. Journal of Contemporary Water Research & Education, 141(1): 45 - 59.

Yeung, C.W., 2005. Rainfall-runoff and water-balance models for management of the Fena Valley reservoir, Guam, Scientific investigations report 2004-5287, U.S. Geological Survey.

List of symbols

Symbol	Description	Unit
A	reservoir surface area	(km^2)
$ALPHA\_BF$	baseflow alpha factor	(day)
$APE\ (j)$	average area per land use type j per unit	(m^2)
$AREA\ (i)$	area of cell i	(m^2)
A_f	wetted area	(m^2)
A_i	area of polygon belongs to rain gauge i	(km^2)
a	wall friction coefficient	(1/m)
$BIOMIX$	biological mixing efficiency	(-)
$BLAI$	leaf area index for crop	(-)
C	Chezy coefficient	($m^{1/2}/s$)
$CANMX$	maximum canopy index	(-)
CH_K2	effective hydraulic conductivity in main channel alluvium	(mm/hr)
CH_N	Manning coefficient for channel	(-)
$CN2$	SCS runoff curve number for moisture condition II	(-)
DAM	direct flood damage	(Thai Baht)
DPE	direct flood damage per land use type	(Thai Baht)
d	depth below plane of reference	(m)
E	reservoir evaporation and other losses	(mm/day)
$EPCO$	plant evaporation compensation factor	(-)
$ESCO$	soil evaporation compensation factor	(-)
E_a	amount of actual total evaporation	(mm, mm/day)
E_{NS}	Nash-Sutcliffe coefficient	(-)
$GWQMN$	threshold depth of water in the shallow aquifer required for return flow to occur	(mm)
GW_DELAY	groundwater delay	(day)
GW_REVAP	groundwater 'revap' coefficient	(-)
g	gravity acceleration = 9.81 (m/s^2)	(m/s^2)
H	maximum flood depth	(cm, m)
h	total water height above the 2D bottom; $\zeta + d$	(m)
h_l	water level with respect to the reference level	(m)
I	tributary inflow to the reservoir	(m^3/day)
L	flood duration	(day)
n	Manning's roughness coefficient	($s/m^{1/3}$)
O	outflow	(m^3/day)
P	rainfall	(mm/day)
P_i	amount of rainfall at rain gauge i	(mm/day)
$PC\ (i,j)$	percentage of land use type j in cell i	(-)
$\overline{P}$	mean areal rainfall over a river basin	(mm/day)
Q	discharge	(m^3/s)
Q_{gw}	amount of groundwater flow	(mm)
Qk_l	1D discharge flowing out of control volume through link k_l	(m^3/s)
Q_{surf}	amount of surface runoff	(mm)

Symbol	Description	Unit
q_{lat}	lateral discharge per unit length	(m^2/s)
R	hydraulic radius	(m)
RCHRG_DP	groundwater recharge to deep aquifer	(fraction)
REVAPMN	threshold depth of water in the shallow aquifer for 'revap' to occur	(mm)
r^2	goodness of fit	(-)
S	water storage	(m^3)
SLOPE	average slope steepness	(m/m)
SLSUBBSN	average slope length	(m)
SOL_ALB	soil albedo	(-)
SOL_AWC	available water capacity of the soil layer	(mm H_2O/mm soil)
SOL_K	soil conductivity	(mm/hr)
SOL_Z	soil depth	(mm)
SURLAG	surface runoff lag coefficient	(-)
SW_0	initial soil water content	(mm)
SW_t	final soil water content	(mm)
s	slope of energy grade line	(-)
s_{ext}	standard deviation of the sample	(mm)
T	return period	(year)
TDM	total direct flood damage	(Thai Baht)
TLAPS	temperature lapse rate	(°C/km)
t	time	(s, day)
u	velocity in x-direction	(m/s)
V	water volume	(m^3)
V	velocity; $V = \sqrt{u^2 + v^2}$	(m/s)
V'	combined 1D/2D volume	(m^3)
v	mean flow velocity	(m/s)
v	velocity in y-direction	(m/s)
W_f	flow width	(m)
w_{seep}	amount of water entering the vadose zone from the soil profile	(mm)
X	unlimited exponentially-distributed variable	(mm)
$\bar{X}_{ext}$	mean of the independent extreme values	(mm)
x	distance in x-direction	(m)
y	distance in y-direction	(m)
y	reduced variate	(1/year)
$\bar{y}_N$	mean of the reduced variate	(1/year)
Δt	time step	(s)
Δx	2D grid size in x (or *i*) direction	(m)
Δy	2D grid size in y (or *j*) direction	(m)
ζ	water level above the plane of reference (the same for 1D and 2D)	(m)
τ_{wi}	wind stress shear	(N/m^2)
ρ_w	water density = 1,000 (kg/m^3) at 4 °C	(kg/m^3)
σ_N	standard deviation of the reduced variate	(1/year)

Acronyms

AFDC	Armed Forces Development Command
ALRO	Agricultural Land Reform Office
BB	Bureau of the Budget
BMP	Best Management Practices
CBO	Community Based Organization
CPD	Cooperative Promotion Department
DDS	Department of Drainage and Sewerage, Bangkok Metropolitan Administration
DEM	Digital Elevation Model
DIW	Department of Industrial Works
DGR	Department of Groundwater Resources
DNP	Department of National Parks, Wildlife and Plant Conservation
DOF	Department of Fisheries
DOPA	Department of Provincial Administration
DPM	Department of Disaster Prevention and Mitigation
DPT	Department of Public Works and Town & Country Planning
DRH	Direct Runoff Hydrograph
DWR	Department of Water Resources
DWSM	Dynamic Watershed Simulation Model
EGAT	Electricity Generating Authority of Thailand
GIS	Geographic Information System
GUI	Graphical User Interface
HBV	Hydrologiska Byråns Vattenbalansavdelning
HDRTN	Hydrographic Department, Royal Thai Navy
HEC-HMS	Hydrologic Modelling System
HRU	Hydrological Response Unit
HSPF	Hydrologic Simulation Program - Fortran
IEAT	Industrial Estate Authority of Thailand
IHMS	Integrated Hydrological Modelling System
LDD	Land Development Department
MAE	Mean Absolute Error
MCM	Million Cubic Metres
MD	Marine Department
MSL	Mean Sea Level
MWA	Metropolitan Waterworks Authority
NESDB	Office of the National Economic and Social Development Board
NGO	Non-Governmental Organization
NHA	National Housing Authority
NRCT	Office of the National Research Council of Thailand
NWRC	National Water Resources Committee
O&M	Operation and Maintenance
OCSC	Office of the Civil Service Commission
ONEP	Office of Natural Resources and Environmental Policy and Planning
OPSMOAC	Office of the Permanent Secretary for Agricultural and Cooperatives
PARASOL	Parameter Solutions Method
PCD	Pollution Control Department

PDF	Probability Density Function
PI	Published Interface
PWA	Provincial Waterworks Authority
RFD	Royal Forest Department
RID	Royal Irrigation Department
RMSE	Root Mean Square Error
RR	Rainfall Runoff
SCS	Soil Conservation Services
SMDR	Soil Moisture Distribution and Routing
SRTM	Shuttle Radar Topographic Mission
SWAT	Soil and Water Assessment Tool
TMD	Thai Meteorological Department
UH	Unit Hydrograph
UHM	Unit Hydrograph Module
USLE	Universal Soil Loss Equation

Appendices

Appendix A. Sensitivity analysis for SWAT model parameters

Sensitivity analysis for hydrological models is a matter of great importance since it is useful in all stages of the modelling process. Proper consideration can speed up the optimization process with a limited number of parameters that influence the model outputs of interest. A description of the 22 parameters considered herein is shown in Table A.1. The order is arranged from the most sensitive to the least sensitive.

Table A.1 List of SWAT input parameters ranks used in sensitivity and calibration analysis

Parameter	Description	Unit	Process	Sensitivity ranking
CN2	SCS runoff curve number for moisture condition II	-	Runoff	1
ALPHA_BF	Baseflow alpha factor	day	Groundwater	2
CH_K2	Effective hydraulic conductivity in main channel alluvium	mm/hr	Channel	3
SURLAG	Surface runoff lag coefficient	-	Runoff	4
ESCO	Soil evaporation compensation factor	-	Evaporation	5
SOL_AWC	Available water capacity of the soil layer	mm H_2O/ mm soil	Soil	6
SOL_Z	Soil depth	mm	Soil	7
SOL_ALB	Soil albedo	-	Evaporation	8
CH_N	Manning coefficient for channel	-	Channel	9
SLOPE	Average slope steepness	m/m	Geomorphology	10
SOL_K	Soil conductivity	mm/hr	Soil	11
SLSUBBSN	Average slope length	m	Geomorphology	12
CANMX	Maximum canopy index	-	Runoff	13
EPCO	Plant evaporation compensation factor	-	Evaporation	14
GWQMN	Threshold depth of water in the shallow aquifer required for return flow to occur	mm	Soil	15

Table A.1 List of SWAT input parameters ranks used in sensitivity and calibration analysis (cont'd)

Parameter	Description	Unit	Process	Sensitivity ranking
RCHRG_DP	Groundwater recharge to deep aquifer	Fraction	Groundwater	16
GW_DELAY	Groundwater delay	day	Groundwater	17
BIOMIX	Biological mixing efficiency	-	Soil	18
GW_REVAP	Groundwater 'revap' coefficient	-	Groundwater	Insignificant
REVAPMN	Threshold depth of water in the shallow aquifer for 'revap' to occur	mm	Groundwater	Insignificant
TLAPS	Temperature lapse rate	°C/km	Geomorphology	Insignificant
BLAI	Leaf area index for crop	-	Crop	Insignificant

Note: Parameters, which were rendered in a bold font, were optimised in the autocalibration. The others were kept constant and taken from the literature/default values.

Appendix B. Goodness-of-fit statistics for numerical and graphical comparison of measured and simulated hydrological time series

In order to evaluate the prediction ability of a mathematical model in the sense that the model behaviour indeed fits well with the observation, it is recommended not to rely merely on one particular statistical measure since it does not provide adequate evidence to describe the performance of a model. Therefore, before making a definite conclusion about the reliability of the model, four statistical measures are used and each of them is briefly described below.

Nash-Sutcliffe coefficient (E_{NS})

The Nash-Sutcliffe coefficient is represented with the following equation:

$$E_{NS} = 1 - \frac{\sum_{i=1}^{n}(O_i - P_i)^2}{\sum_{i=1}^{n}(O_i - O_{avg})^2} \tag{B.1}$$

where (note: if a variable changes, it causes a change in unit):

O_i = observed value at time step i

P_i = simulated value at time step i

O_{avg} = average observed value of the simulation period

n = number of observations

In the presence of this coefficient, the value 1 indicates perfect model prediction, whereas the value of 0 implies unacceptable performance.

Root Mean Square Error (RMSE)

The Root Mean Square Error is defined as follows:

$$RMSE = \left[\frac{1}{n}\sum_{i=1}^{n}(O_i - P_i)^2\right]^{1/2} \tag{B.2}$$

The smaller the *RMSE* value, the better the model performance.

Goodness of fit (r^2)

The goodness of fit of the models is calculated using the following formula:

$$r^2 = \left[\frac{\sum_{i=1}^{n}(O_i - O_{avg})(P_i - P_{avg})}{\left(\sqrt{\sum_{i=1}^{n}(O_i - O_{avg})^2}\right)\left(\sqrt{\sum_{i=1}^{n}(P_i - P_{avg})^2}\right)}\right]^2 \tag{B.3}$$

where (note: if a variable changes, it causes a change in unit):
P_{avg} = average simulated value of the simulation period

The r^2 value varies from 0 to 1, with 1 means a good fit and 0 would imply that the model fails to perform accurately.

Mean Absolute Error (MAE)

The Mean Absolute Error is determined by:

$$MAE = \frac{1}{n}\sum_{i=1}^{n}|O_i - P_i| \quad (B.4)$$

If this value is zero, the model prediction ability is perfect.

Appendix C. Schematic diagram of 1D/2D SOBEK hydraulic model

To simplify the Chi River hydraulic system during the model simulation, the actual components of a river system are represented in the form of schematic diagram. It is composed of the essential network and connectivity through nodes and links, which interconnect to form a river system. From this connectivity, the Chi river network schematization representing those elements is shown in Figure C.1 and C.2.

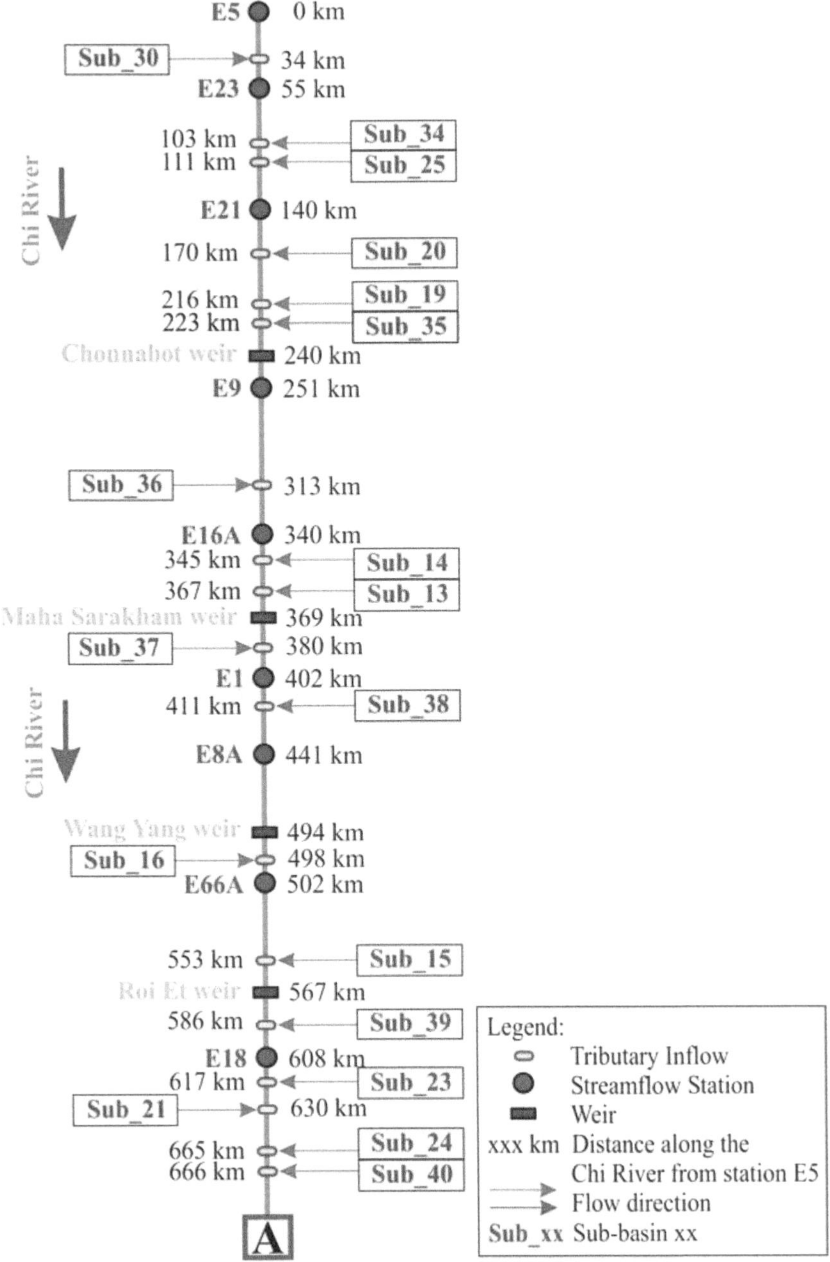

Figure C.1 Schematic illustration of the Chi River network and its tributaries

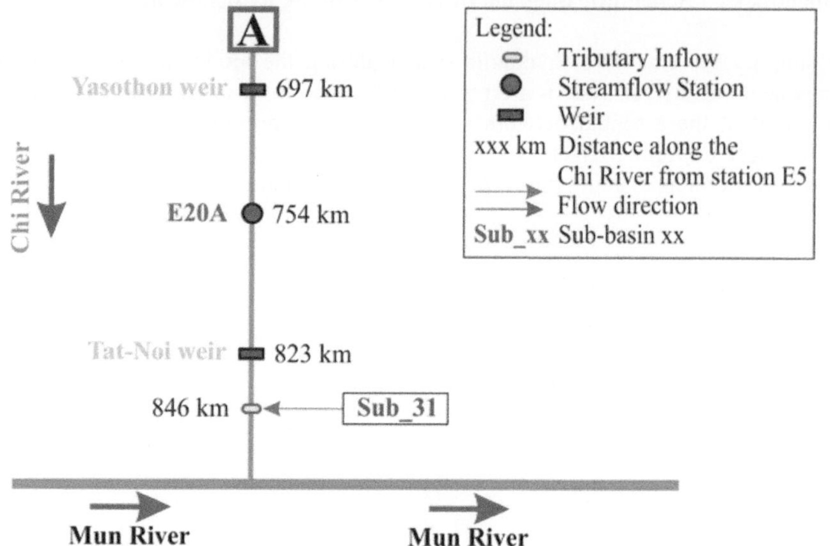

Figure C.1 Schematic illustration of the Chi River network and its tributaries (cont'd)

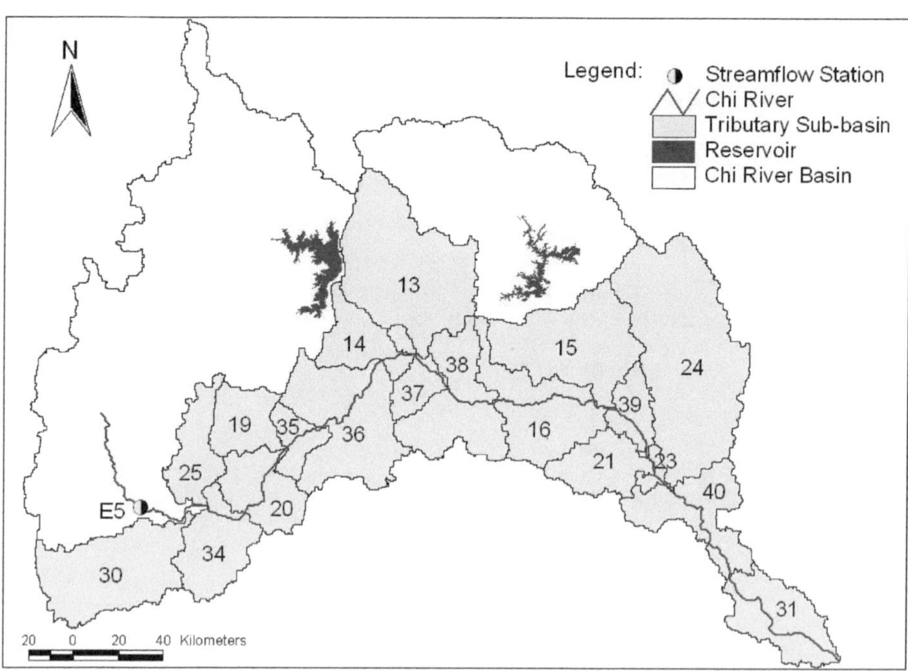

Figure C.2 Main stem and individual tributary sub-basins identified areas

Appendix D. Detailed descriptions of weir structures on the Chi River

There is a growing recognition towards the completion of weir construction, it highlights the fact that weir structures on the Chi River are indeed likely to disrupt the natural flow of water. They are usually used to store water for irrigation in the dry season in order to meet the requirements for each particular irrigated area. Moreover, they are always opened to their full extent during the flood season. A row of different number of weir openings of each site equipped with radial gates. The weir locations and their profile are presented in Figure D.1 and D.2, while the characteristics of the weir gates are shown in Table D.1.

In order to bring the model behaviour closer to the actual physical behaviour, the flow in 1D river network, which was connected to 2D grid cells, was modelled with extensive details of hydraulic principles of weir behaviour within the 1D system.

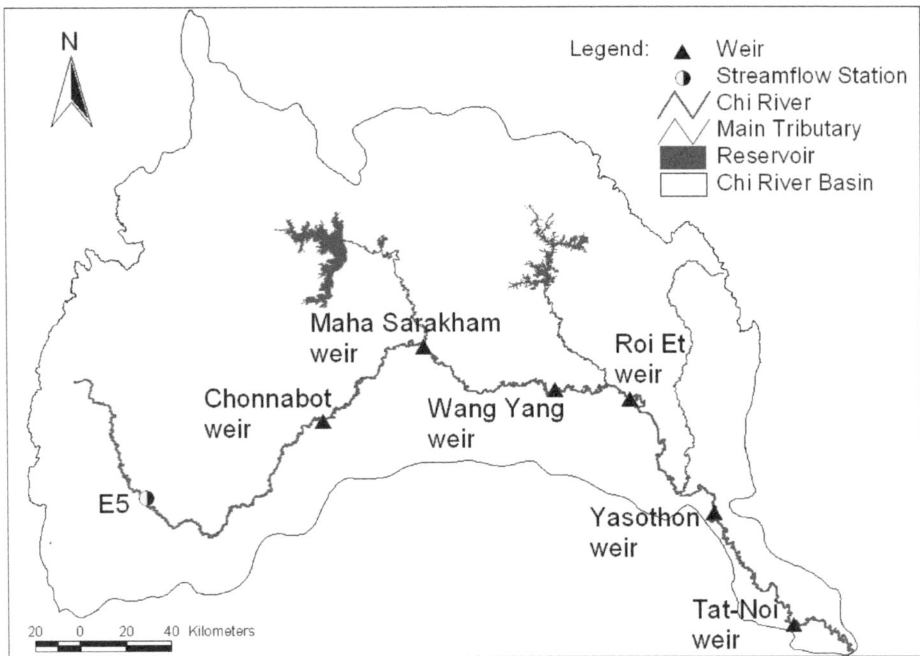

Figure D.1 Locations of weirs on the Chi River

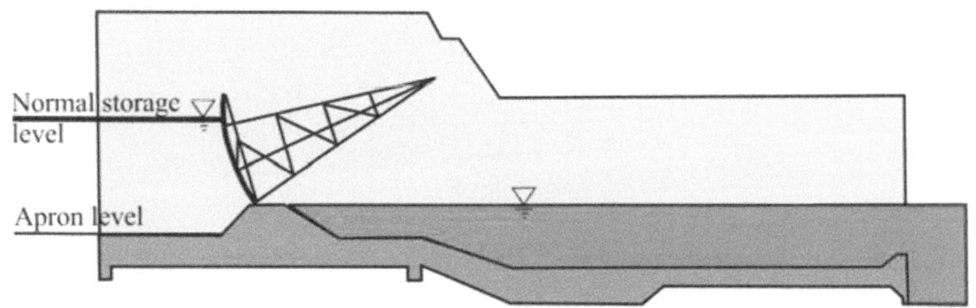

Figure D.2 Longitudinal section through a weir in closed position of all six weirs along the Chi River

Table D.1 Individual gate dimensions of weir gates

Name	Specifications				
	No. of gates	Gate width in m	Gate height in m	Apron level in m+MSL	Normal storage level in m+MSL
Chonnabot	6	10.0	8.60	152	162.0
Maha Sarakham	6	12.0	7.00	138	146.8
Wang Yang	6	12.5	6.60	129	137.0
Roi Et	6	12.5	7.75	124	130.0
Yasothon	8	12.5	7.75	118	126.0
Tat-Noi	8	12.5	7.50	107	116.0

Note: MSL = Mean Sea Level

Appendix E. Potential benefits and shortcomings of flood mitigation measures

Prior to the investigation, it is crucial to take notes from overall review of various possible actions for managing floods. These cover three flood mitigation categories such as flood modification (keeping floods away from people and property), property modification (keeping people and property away from floods) and response modification (modifying human behaviour through activities) (note: some of the tasks among them are complimentary to one another). A more detailed discussion of their potential advantages and disadvantages is outlined in Table E.1.

Table E.1 Summary of the potential advantages and disadvantages of flood management measures

Measure	Advantage	Disadvantage
Flood modification		
1) Channel modification[1]	- more floodwater is carried away at a faster rate; - provides site-specific protection; - no land requirements; - effects of construction are localized.	- can create/worsen flooding problems downstream; - high cost of maintenance to preserve the increased conveyance capacity created; - the potential to cause problems of erosion and sedimentation, which can result in changing river morphology.
2) Flood mitigation reservoir[1]	- decreases the flood peak and provides the massive protection in the area immediately downstream of the reservoir.	- high construction costs; - can create false sense of security within downstream residents since the reservoir only controls floods up to some particular size for which it was designed; - requires large areas and possible displacement of large populations; - protection may decrease in time if the reservoir capacity is diminished by siltation.
3) Bypass/flood diversion[1]	- increases flood capacity of a system by creating an alternative overflow or storage channel for floodwaters that can no longer be carried within the river channel; - improves flow characteristics and decreases flood levels along the channel downstream; - serves other functions, such as providing additional farmland or parkland, when not needed to convey floodwaters.	- in built-up areas, availability of low-value land can be limited and expensive; - may cause environmental and morphological impacts; - may transfer flood problem to other areas; - opportunities for constructing are limited by the topography of the area; - very costly in terms of construction and maintenance.
4) Retention basin[1]	- controls flooding by temporarily storing floodwaters. After a flood peaks, water is released slowly at a rate that the river can accommodate downstream;	- area suffers damage when flooded; - local residents can be at risk when the impact on floods larger than those for which it was designed;

Table E.1 Summary of the potential advantages and disadvantages of flood management measures (cont'd)

Measure	Advantage	Disadvantage
4) Retention basin[1] (cont'd)	- reduces downstream flood peak and confine flooding to within the flood control system; - suitable for protecting existing development downstream from the project site; - can be used for productive agricultural purposes during dry periods.	- effectiveness depends on its location, available storage capacity and residual drainage facilities; - can negatively impact downstream flows due to release timing and rate; - high construction costs; - requires a substantial area to achieve the necessary storage; - suitable geologically and topographically sites may require very considerable and expensive land acquisition; - long duration or multi-peak storms (when the basin is filled from a previous peak) can increase the risk of overtopping or breaching.
5) Dike[1]	- confines floodwater in the river channel for either a selected portion of the floodplain or a larger area; - useful measure against flooding up to the design level; - permits further development of existing urban areas.	- requires large space; - may push floodwater to other areas and raise flood levels elsewhere on the floodplain; - high construction and maintenance costs; - can cause disastrous damages if overtopping does occur; - may create a false sense of security about the degree of protection provided; - can impact environment and scenic views; - complicates the drainage of land they protect; - protects only the area immediately behind it, and is effective only against flood depths up to the chosen level for which it was designed.
Property modification		
6) Land use and spatial planning[2]	- limits potential for increased damages to future development in flood prone areas; - permitted land use is consistent with flood hazard; - ensures that the flood problem is not worsened by new structure in flooded areas; - environmental character of the area is largely preserved.	- can restrict legitimate development if controls are not consistent with flood hazard; - controls can be too restrictive; - conflicting interests of assessment through development and flood protection can cause problems; - will not impact the high level of flood damages to existing development, but the future.

Table E.1 Summary of the potential advantages and disadvantages of flood management measures (cont'd)

Measure	Advantage	Disadvantage
7) Property acquisition and floodway clearance[2]	- can lead to the free flow of floodwater and reduces flood levels; - vulnerable development is removed from flood hazard areas; - reduces post-flood rehabilitation costs.	- may not be readily accepted by residents in the area; - acquisition costs can be high if carried too far; - needs to be undertaken in conjunction with a re-settlement programme if squatter populations are involved.
8) Flood proofing[2]	- reduces post-flood clean-up operations; - especially suits to commercial and industrial buildings; - ensures structures in the flood prone areas will be flood-proofed; - can reduce the increase in potential flood damage.	- suits only to particular types of buildings; - can result in high losses if flood height is exceeded; - uniform application is difficult and may well be unacceptable.
9) Flood fighting[2]	- reduces adverse impacts of flooding including casualties and damages.	- requires effective flood warning system; - requires detailed planning and trained personnel; - high cost to government.
Response modification		
10) Public information, education and involvement[2]	- creates an awareness of flood hazard confronting communities; - generates more ready acceptance of flood loss prevention measures; - affected person may provide useful information that cannot easily be quantified; - helps to enhance public confidence.	- can be a time consuming task for flood authority; - can be unproductive if community acceptance is negative; - there is a potential for confusion of the erroneous information.
11) Flood forecasting and warning systems[2]	- minimizes damage and danger to life; - alerts general public to take actions; - generally cheap and quick to install; - useful in conjunction with other measures.	- feasible only for river basins with long response times; - limited usefulness for small river basins; - tends to be ignored if constantly inaccurate; - needs to be integrated with other measures.
12) Evacuation from endangered areas[2]	- reduces loss of life; - generally cheap to introduce.	- requires effective flood warning system; - requires detailed planning; - public awareness must be maintained; - it is sometimes not considered to be economically or politically feasible, and is not publicly acceptable.

Table E.1 Summary of the potential advantages and disadvantages of flood management measures (cont'd)

Measure	Advantage	Disadvantage
13) Flood adaptation[2]	- reduces potential flood losses to individuals; - individuals meet their own costs.	- only applicable in non-hazardous areas; - may not provide protection against major flood conditions.
Others		
14) Flood relief[2]	- reduces financial burden on affected individuals; - reduces impact of post flood problems.	- funds are provided by entire community; - generous funds and assistance tend to promote continued occupation of hazardous areas.
15) Flood insurance[2]	- can provide immediate financial relief to cover individual flood loss or damage; - can reduce the demand on flood relief funds provided to victims; - encourages and places responsibility on individuals who choose to reside in flood hazard areas, for their future flood losses.	- it is sometimes hard to obtain from private insurance companies; - can promote development in floodplain; - can leave people uninsured or underinsured; - national flood insurance schemes require use of public funds; - can only assure for financial losses suffered during a flood and does not reduce flood losses to either existing or future development.

Note: [1] Considered as an alternative to structural measures
[2] Considered as an alternative to non-structural measures

Sources: Economic and Social Commission for Asia and the Pacific (1991, 2007)
International Commission on Irrigation and Drainage (1999)
van Duivendijk (2005)
Douben (2006)

Appendix F. A design criterion for upstream flood retention basin

After a site selection of a retention basin has been made in which its consequences were already known and expected, the upstream flood retention storage is assigned to serve as a temporary storage of diverted water from the Chi River. Then, the immense flood storage potential will be of interest. To provide some context to this, its success would have to be compatible with a specially designed outlet and operation, which is very important and is able to specify how the assigned retention basin meets optimal flood control requirements.

During a 1% annual probability of exceedance flood event, flap gate on the intake side is designed to control/adjust flow into the retention basin via flood bypass (trapezoidal in shape) when the Chi River is approaching a flood stage. The gate opening and closing operation is formed by the pressure of the water being impounded. Its basic design is shown in Figure F.1. In the meantime, the outflow control structure remains closed in order to prevent floodwater from discharging until additional space becomes available in the Chi River. While the Chi water level is below 170 m+MSL, the captured floodwater in the retention basin can flow through another trapezoidal-shaped flood diversion channel (Figure F.2).

Figure F.1 Moveable flap gate at the inlet of the proposed upstream flood retention storage

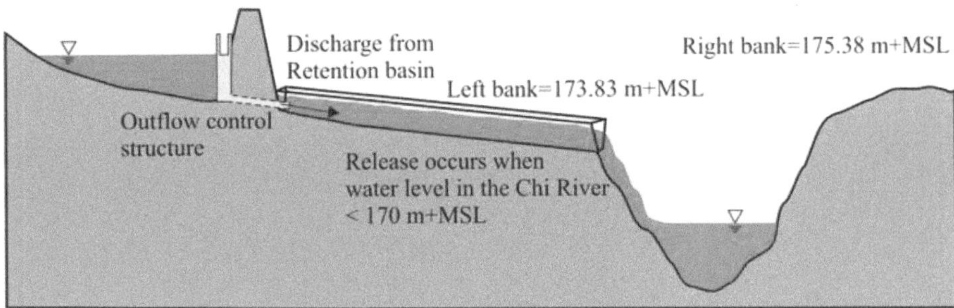

Figure F.2 Schematic drawing of the release approach from the retention basin into the Chi River

Samenvatting

Overstromingen behoren tot een van de grootste natuurrampen die leiden tot verlies aan mensenlevens en die van invloed zijn op de sociale en economische ontwikkeling. In veel gevallen heeft de natuur laten zien weinig respect te hebben voor onverstandige beperking van haar recht op vrije doorgang door de mens met betrekking tot verstedelijking en industrialisatie door overstroming van menselijke bezittingen en het wegnemen van levens. Overstromingen kunnen echter ook belangrijke ecologische voordelen brengen voor riviersystemen. Op dit moment lijkt het erop dat zich meer en meer potentiële overstromingsgebieden ontwikkelen en doet zich de volgende vraag voor: Hoeveel van die ernstige overstroming bedreigingen worden eigenlijk veroorzaakt door verschillende menselijke activiteiten? Vaak zijn de oorzaken van de overstromingen:

- ontbossing in een stroomgebied;
- veranderingen in grondgebruik die leiden tot veranderingen in de waterhuishoudkundige dynamiek;
- verstedelijking en bevolkingsgroei in gebieden waar overstromingen kunnen optreden;
- rivier regelgeving, aanleggen van dijken en kanalisatie;
- verhoging van waterstanden ten gevolge van natuurlijke of kunstmatige obstakels in het rivierbed en het overstroombare gebied, bijvoorbeeld bruggen en stuwen;
- falen van dammen en dijken.

Over het algemeen zijn er twee belangrijke manieren waarop mensen hebben geprobeerd om overstromingen te beheersen:

- *structurele maatregelen:* dammen, dijken, aanpassingen in het rivierbed en het overstroombare gebied, bypasses, enz., die zijn ontworpen om het optreden, of de omvang van overstromingen te beperken. Dat wil zeggen controle van hoogwater door opslag, inperking, wijziging van de stroming, of afleiding;
- *niet-structurele maatregelen:* het voorspellen van overstroming, waarschuwing voor overstroming, controle van ontwikkelingen in het rivierbed en het overstroombare gebied, verzekering tegen overstroming, evacuatie, enz., die zijn gepland om nadelige effecten van overstromingen te elimineren of te beperken.

In het verleden kregen structurele maatregelen tegen overstroming prioriteit. Echter, in de afgelopen jaren wordt meer de nadruk gelegd op niet-structurele maatregelen om de kwetsbaarheid en de enorme economische, sociale en menselijke verliezen te beperken en tegelijkertijd de ecologische gevolgen van mitigerende maatregelen te minimaliseren.

Het stroomgebied van de Chi rivier ligt in het noordoosten van Thailand. De totale oppervlakte bedraagt 4,9 miljoen ha met een bevolking van 6,6 miljoen mensen. Het is gelegen in de tropische moesson regio. De jaarlijkse neerslag varieert van 1.000 - 1.400 mm. De Chi rivier is de belangrijkste rivier en is tevens de langste rivier in Thailand met een totale lengte van 946 km. De fysieke kenmerken van het stroomgebied van de Chi rivier vertonen sterke verschillen, met steile hellingen in de bovenstroomse gebieden en vlakke laaggelegen gebieden, dat wil zeggen de brede, vlakke overstroombare gebieden in de benedenloop, vooral in de buurt van de samenvloeiing met de Mun rivier, een belangrijke zijrivier van de Mekong rivier.

In het kader van demografische dynamiek, is de bevolkingsgroei in het stroomgebied van de Chi rivier sinds 1993 toegenomen tot ongeveer 0,6% per jaar. Het grootste deel van de groei heeft zich voorgedaan op het platteland, waar ongeveer 61% van de mensen leven. Al deze effecten, hebben op hun beurt belangrijke implicaties voor de ruimtelijke dynamiek, die direct van invloed zijn de vraag naar land en het gebruik ervan. Momenteel is de meerderheid van de bevolking in het stroomgebied werkzaam in de landbouw op de 60% akkerland, waarvan 41% is rijstvelden. Het resterende gebied betreft bos (31%), stedelijk gebied (2,9%), waterlopen en meren (2,5%), en overig grondgebruik (3,5%).

Met betrekking tot overstromingen, heeft het stroomgebied van de Chi rivier in de afgelopen decennia last gehad van een aantal grote overstromingen met dodelijke slachtoffers, verplaatsing van mensen, uitgebreide fysieke schade samen met enorme economische verliezen en grote effecten op de natuur. Het is duidelijk dat de bestaande maatregelen ter beperking van overstroming geen aanzienlijk effect hebben gehad in termen van vermindering van de overstroming en de totale schade erdoor. Maatregelen tegen overstromingen in het stroomgebied van de Chi rivier zijn voornamelijk beperkt tot het bouwen van dijken die over het algemeen zijn aangelegd om stedelijke gebieden of landbouwgrond te beschermen tegen relatief frequente overstromingen, vooral in het benedenstroomse deel van het stroomgebied. Deze voorzieningen ter verdediging, gebouwd door het Koninklijke Departement voor Irrigatie in de jaren 1950, 1970, en 1980, zijn meestal ontworpen om bescherming te bieden tegen een overstroming met een kans van overschrijding van 10% per jaar.

Het belangrijkste doel van dit onderzoek is om een geïntegreerd modellen raamwerk te ontwikkelen dat een meer holistische benadering voor overstromingsbeheer, in relatie tot het grondgebruikgebruik en de veranderingen in het stroomgebied van de Chi rivier mogelijk maakt. Er is een groeiend besef dat met een geïsoleerde aanpak van het beheer van overstromingen een zekere mate van beperken van overstromingen in een bepaald gebied kan worden bereikt. Dit zou echter niet de optimale oplossing kunnen zijn om toekomstige overstromingen rampen te beperken. Daarom is een geïntegreerde en geoptimaliseerde aanpak voor het treffen van overstromingen mitigerende maatregelen, dat wil zeggen niet een aantal stand-alone maatregelen, vereist. Een meer geïntegreerde aanpak van de beheersing van de gevolgen van overstromingen is nodig om passende maatregelen, dat wil zeggen met zorg geselecteerde punten en soorten van interventies, te nemen. Een geïntegreerde aanpak van waterveiligheid zal leiden tot de beste combinatie van structurele en niet-structurele maatregelen.

Modellering benaderingen van overstroming effectbeoordeling

Een overstroming is een complexe hydrologische gebeurtenis. Modellen kunnen niet alleen helpen om dit fenomeen beter te begrijpen, maar zijn essentieel voor beoordeling van overstromingsrisico's in de huidige situatie en om de effecten van de voorgestelde wijzigingen in overstromingsgevoelige gebieden te beoordelen.

Hydrologische modellering

Toepassing van hydrologische modellen is een geschikte manier om bij te dragen aan een beter begrip van de hydrologische processen en het synthetiseren van de beschikbare gegevens (tijdreeksen en GIS-gegevens) in het stroomgebied van de Chi rivier. Echter, wanneer men wordt geconfronteerd met dit vereiste, rijst natuurlijk de vraag welk model het meest geschikt is voor dit specifieke onderzoek. Om deze vraag te beantwoorden is het op het proces gebaseerde hydrologisch model, SWAT (Bodem en Water Assessment Tool), dat is een goed ontwikkeld, veel gebruikt en robuust model, gebaseerd op dagelijkse tijd stappen, gekozen.

Het SWAT model is het meest geschikt voor landelijke gebieden met overwegend agrarisch grondgebruik, wat inderdaad overeenkomt met de omstandigheden in het stroomgebied van de Chi rivier. Het biedt ook de mogelijkheid voor gedetailleerde invoer van gegevens betreffende veranderingen in de vegetatie en landbouwpraktijken. Het model kan grote hoeveelheden gegevens geïntegreerd verwerken. Bovendien werd SWAT geselecteerd om gebruikt te worden in deze studie, omdat het een vrij beschikbaar model is, samen met een aantal andere voordelen zoals het gratis beschikbaar zijn en de mogelijkheid om het model aan te passen aan de specifieke behoeften. Het model wint aan populariteit onder de gebruikers en is wereldwijd met succes getest onder verschillende geografische en klimatologische omstandigheden. Daarom werd het SWAT model geselecteerd om de effecten van ontwikkelingsactiviteiten in het bovenstroomse deel van het stroomgebied te onderzoeken en het stroomgebied te modelleren om de veranderingen door maatregelen ter beperking van overstromingen te analyseren.

Het SWAT model is getest of het geschikt is om instroom vanuit deelstroomgebieden door zijrivieren goed te identificeren en te kwantificeren. De SWAT simulaties op basis van dagcijfers hadden betrekking op de periode 1 januari 2000 tot en met 31 december 2001. De gekozen periode voor de simulatie is gebaseerd op historische gegevens, omdat in deze periode een van de meest verwoestende overstromingen in de geschiedenis van het stroomgebied van de Chi rivier is opgetreden. Een initialisatie periode van 1,5 jaar is gebruikt, zodat de gevolgen van onzekere initiële condities in het model werden geminimaliseerd. Omdat het SWAT model vele parameters kent, die in de regel niet waar te nemen zijn, moet het model worden gekalibreerd op gemeten gegevens om de beste of op zijn minst een redelijke set parameter waarden te bepalen. Bovendien, omdat er geen tijdreeksen met dagcijfers beschikbaar zijn van de afvoer van het stroomgebied van de Chi rivier, is de kalibratie uitgevoerd voor de ter plaatse van Ban Tha Khrai (station E18) gemeten waarden binnen het stroomgebied. Tijdens het kalibratie proces is een gevoeligheidsanalyse uitgevoerd om de zes parameters waarvoor de model resultaten het meest gevoelig zijn te bepalen. Vervolgens zijn deze zes parameters gebruikt om het model voor de periode 1 juni - 31 oktober 2001 te kalibreren.

Met betrekking tot de kalibratie resultaten op afvoermetingen voor het station E18, zijn diverse statistische toetsen toegepast om de simulatie nauwkeurigheid te evalueren, zoals de Nash-Sutcliffe coëfficiënt, wortel uit het kwadraat van de gemiddelde fout (Root Mean Square Error), juistheid van de benadering (Goodness of Fit) en gemiddelde absolute fout (Mean Absolute Error). Toen de gegevens over 2002 werden gebruikt als validatie set, bleek dat validatie resultaten opmerkelijk goed waren. De Nash-Sutcliffe coëfficiënt en waarden voor de juistheid van de benadering waren hoog (groter dan 0,85), wat aangeeft dat de prestaties van het SWAT model bevredigend zijn en dat het model in staat is om de afvoer redelijk goed te simuleren. Daarna zijn de gekalibreerde model parameters gebruikt om afvoeren behorend bij verschillende regenval scenario's voor de voorspellingen van de instroom vanuit de zijrivieren met een bepaalde kans van overschrijding op geselecteerde punten van de Chi rivier te simuleren. De instroom vanuit de zijrivieren is toen gebruikt als een belangrijke invoer in het hydraulische model.

Hydraulische modellering

Het vastleggen van de complexe interacties tussen stroming in de rivier, het genereren van afvoer en het gebied dat gevoelig is voor overstromingen is de belangrijkste onderdeel voor het simuleren van de overstroming processen in het overstroombare gebied. Om inzicht te krijgen in het hydraulische systeem moeten de dominante processen daarom goed worden onderkend. Om het complexe dynamische gedrag te adresseren, kan het gebruik van een hydraulisch model het mogelijk maken om complexe stromingspatronen te simuleren.

Een pragmatische aanpak voor het behandelen van overstromingsproblemen in het stroomgebied van de Chi rivier was nodig. Daarom is in deze studie een aanzienlijke inspanning verricht met betrekking tot het toepassen van een geschikt hydraulisch model, dat wil zeggen 1D/2D SOBEK, voor de (integrale) simulatie van overstromingsprocessen. Het gebruikte 1D/2D SOBEK model is gekozen voor het modelleren vanwege het vermogen overstromingen te simuleren, omdat het dynamisch 1D knooppunten kan verbinden met 2D-cellen in de module voor het overstroombare gebied. Bovendien, biedt het model ook veel mogelijkheden voor koppeling met andere modellen en daarmee de mogelijkheid om belangrijke vraagstukken op het gebied van overstromingsbeheer aan te pakken en te verbeteren.

Model kalibratie met de juiste gegevens vormt een cruciale stap in de algehele beschrijving van een valide en betrouwbaar proces. De primaire kalibratie parameter voor het 1D/2D SOBEK model is de Manning ruwheid coëfficiënt (n) van het rivier bed, die kan worden bepaald uit het kalibratie proces met behulp van 1D modellering (SOBEK 1D). Het model is gekalibreerd met waargenomen waterstanden voor de jaren 2000 - 2001 en binnen redelijke grenzen aangepast totdat het model gemeten waterpeil profielen voor meetpunten in het benedenstroomse deel van de Chi rivier op een acceptabele wijze reproduceerde. Vanwege het feit dat de Manning n waarden tussen de locaties kunnen variëren, heeft de kalibratie zich gericht op drie sets van Manning n waarden. De optimale waarden werden gevonden door het vergelijken van de waargenomen en gesimuleerde waterstanden op de Chi rivier in de waarnemingspunten Ban Tha Khrai (E18) en Chana Maha Chai (E20A). De vorm van de hydrographs en het optreden van de pieken correspondeerde goed met de waarnemingen, waaruit kon worden afgeleid dat het model de dynamiek van de piek afvoeren goed weergaf, zoals aangegeven door de hoge waarden van de parameters voor de juistheid van de benadering, dat wil zeggen 0,94 op E18 en 0,93 bij E20A. De resultaten van de validatie (jaar 2000) zijn bijna even goed, dat wil zeggen 0,79 bij E18 en 0,89 bij E20A. Als men rekening houdt met het criterium voor de juistheid van de benadering, geven de aangepaste coëfficiënten voor de Manning ruwheid, die varieerden van 0,028 $s/m^{1/3}$ in de buurt van de monding tot 0,045 $s/m^{1/3}$ in de meer stroomopwaarts gelegen secties, over het gehele traject een positief resultaat van de kalibratie. De waarden van Manning's n van 0,045 $s/m^{1/3}$ voor de sectie vanaf het waarnemingspunt Ban Non Puai (E5) tot de Chi-Nam Phong samenvloeiing, 0,032 $s/m^{1/3}$ van de Chi-Nam Phong samenvloeiing tot de Chi- Nam Yang samenvloeiing, en 0,028 $s/m^{1/3}$ van de Chi-Yang Nam samenvloeiing tot de Chi-Mun samenvloeiing bleken bevredigende resultaten van de kalibratie te geven en correspondeerden met redelijke waarden voor Manning's n volgens de tabel van Chow. Er kon echter geen verdere kalibratie worden uitgevoerd voor de hydraulische ruwheid voor stroming over het land (2D SOBEK). De parameter is daarom afgeleid uit de literatuur (0,1 $s/m^{1/3}$). Omdat geen gedetailleerde informatie betreffende verschillen in ruimtelijke ruwheid beschikbaar was is uitgegaan van een constante waarde over het overstroombare gebied.

Geïntegreerde hydrologische en hydraulische modellering

Het is duidelijk dat de hydrologische en hydraulische processen interactief moeten worden gemodelleerd door middel van een volledig geïntegreerde modellering door het ontwikkelen van een robuust en hydrologisch goed pakket van algoritmen die (volledig) zijn geïntegreerd onder een gebruiksvriendelijke interface. Als gevolg hiervan is het optimale model aangepast door een seriële koppeling van de juiste modules/componenten voor de gewenste toepassing. Hoewel de koppeling van de modellen nodig is, kan het om vele redenen behoorlijk ingewikkeld zijn om het te implementeren, dat wil zeggen formateren van gegevens, compatibiliteit van schalen, enz. Echter, de toegankelijkheid van het Delft-FEWS platform, software ontwikkeld door Deltares, biedt de mogelijkheid van volledige koppeling en uitwisseling van gegevens tussen hydrologische en hydraulische modellen. Dit maakt het mogelijk om de complexe interacties vast te leggen, maar toch relatief eenvoudig toe te passen voor verschillende toepassingen. Bovendien is Delft-FEWS in staat om de modellen te integreren in een handig pakket voor modellering van overstromingen, interactie tussen de twee modellen is een zeer efficiënte manier om de veelzijdigheid van dit instrument voor de overstroming studies te tonen.

In deze studie is het hydrologische neerslag-afvoer model gebruikt om de tweede ronde die voortvloeien uit twee 'kritische gebeurtenissen', de eerste op basis van de vloed van 2001, die overeenkomt met een overstroming met een kans van overschrijding van 4% per jaar en de tweede op basis van extreme neerslag resulterend in een overstroming met een kans van overschrijding van 1% per jaar. De hydrologische simulaties resulteren in invoer voor het erop volgende hydraulische model om overstroming in het overstroombare gebied te simuleren met betrekking tot de schadelijke gevolgen van mogelijke overstromingen. De koppeling tussen de twee modellen werd uitgevoerd in een multi-scenario overstroming modellering experiment, te weten het analyseren van toekomstige veranderingen in de hydrologische en aanverwante systemen, grondgebruik, enz.

De koppeling tussen SWAT en 1D/2D SOBEK is gerealiseerd door een reeks knooppunten waar de zijrivieren afvoeren op de Chi rivier, hierbij is aangenomen dat er geen directe terugkoppeling was van de stroming over het land naar de neerslag-afvoer reactie. Het gekoppelde SWAT-1D/2D SOBEK model werkt als volgt:

- de neerslag-afvoer module (SWAT) vormt het startpunt van het gekoppelde model om een realistische weergave te leveren van de terrestische hydrologische systemen omdat het een schakel vormt tussen de drie belangrijkste hydrologische compartimenten: atmosfeer, oppervlaktewater en grondwater. De hydrologische invoer bepaalt de omvang van de totale extreme afvoer van de verschillende deelstroomgebieden;
- daarna, vormt de afvoer van de deelstroomgebieden de grenzen voor de toevoer van de module voor de stroming over het land (1D/2D SOBEK) op bepaalde locaties door deze te koppelen aan de knooppunten in het netwerk van de rivier. Bij deze stap zijn de SWAT simulaties voor rivierafvoer in de hoofdtak van de Chi rivier uitgeschakeld. Vervolgens is de daaruit voortvloeiende voortplanting van de overstroming op basis van tijdstap van een half-uur gesimuleerd, ondanks dat de afvoer gegevens van de SWAT simulaties op basis van dagcijfers waqren bepaald. Dit omdt het 1D/2D SOBEK model in staat is om invoer te interpoleren naar kleinere tijdsintervallen. De manier van interpoleren kan een belangrijke invloed hebben op de prestaties van het model. Een grote fout in het volume treedt op als de tijdstap te groot is voor de te interpoleren invoer gegevens wanneer deze plotseling

veranderen, dat wil zeggen met een grotere piek-piek variatie op elk tijdstip en bij het veroorzaken van numerieke problemen. Echter, door de relatief kleine tijdstap die is gebruikt in 1D/2D SOBEK kon geen significant effect op de hydraulische simulatie worden gevonden.

De combinatie van de twee veel gebruikte modellen SWAT en 1D/2D SOBEK is in staat om de processen die overstromingen genereren realistisch te simuleren en om de effecten van verschillende scenario's voor het mitigeren van overstromingen te benaderen, dat wil zeggen zowel structurele en niet-structurele maatregelen. Het betreft verschillende factoren die verband houden met overstromingen, zoals verhoogde afstroming volumes en flashiness, het ophouden van afvoer, enz. Los van het eindresultaat heeft model koppeling bewezen dat dit haalbaar is en efficiënt kan zijn en het lijkt een veelbelovende aanpak met grote voordelen voor overstromingsbeheer in het stroomgebied van de Chi rivier. Voorts, hoewel ontworpen voor het stroomgebied van de Chi rivier, kan de model koppeling ook gebruikt worden als een prototype voor model toepassingen in andere gebieden van het land.

Inzicht in de invloed en effecten van de onzekerheid in de context van overstromingsrisico modellering is cruciaal voor het beheer van overstromingsrisico's. Om het proces van het modelleren van hoogwater te verbeteren, is het cruciaal om een beter inzicht te hebben in de onzekerheden die inherent zijn aan het modelleren van processen, en hoe de onzekerheid expliciet is vertaald in de model resultaten die kunnen worden verwerkt in verdere studies.

Analyse van alternatieve overstroming mitigerende maatregelen

De kosten en baten van een bepaalde set van maatregelen zijn op globaal niveau geanalyseerd. Om een alternatieve maatregel voor het mitigeren van overstromingen te rechtvaardigen, zou de reductie van de overstromingsschade groter moeten zijn dan de uitvoeringskosten. Bij het vergelijken van alternatieven zouden de netto voordelen, of wel totale baten minus de totale kosten, groter moeten zijn dan in andere alternatieven waarmee dezelfde reductie in de overstroming te bereiken is.

Kwantificeren van overstromingsschade

De schade veroorzaakt door overstromingen kan worden benaderd als een functie van de kenmerken van de vloed, te weten de diepte en de duur van de overstroming, als gevolg van fysiek contact met vloedwater per categorie van het element in gevaar. In Thailand zijn structurele maatregelen om de gevolgen van overstromingen te weerstaan ontworpen op een gebeurtenis met een bepaalde kans van overschrijding. Het Koninklijke Departement van Irrigatie heeft de overschrijding frequentie (veiligheidsnorm) vastgesteld op een overstroming met een kans van overschrijding van 1% per jaar op een bepaalde locatie waar veel mensen zouden kunnen worden getroffen. Daarom is in deze studie de mogelijke schade beoordeeld op basis van de berekende overstromingsdiepte met een kans van overschrijding van 1% per jaar om de kwetsbaarheid voor overstroming te indiceren en de ruimtelijke verdeling van de potentiële schade in het stroomgebied van de Chi rivier te bepalen. In verband hiermee zijn de van belang zijnde waarden bepaald om de directe voordelen van overstromingen mitigerende maatregelen te schatten in termen van overstromingsschade reductie. Merk op dat gevolgen, zoals de menselijke gezondheid, schade aan het milieu, voordelen of andere indirecte kosten in deze studie niet zijn meegenomen.

Ruimtelijke analyse technieken maken integratie van overstromingsdiepte en grondgebruik mogelijk, teneinde te evalueren welke elementen of eigendommen worden getroffen door een overstroming met een kans van overschrijding van 1% per jaar, alsmede hoeveel ze worden getroffen in termen van overstromingsdiepte. De volgende categorieën van grondgebruik zijn beschouwd in het schade-onderzoek: woongebieden, commerciële en industriële gebieden, landbouw en infrastructuur.

Schade functies zijn bepaald voor de kwantificering van de verschillende categorieën schade in monetaire termen. Op basis van grondgebruik, waarde van de bezittingen en schade functies, is de directe schade als gevolg van een overstroming met een kans van overschrijding van 1% per jaar bepaald. Met de directe schade aan de infrastructuur is echter geen rekening gehouden. Daarom is in deze studie, de schade aan de infrastructuur geschat als een vaste 65% fractie van de andere schade door overstroming. Met behulp van deze schade functies, is de schade aan verschillende vormen van grondgebruik geraamd en de som van de totale directe schade door overstromingen.

Geraamde kosten voor alternatieve maatregelen op het gebied van hoogwaterbeheersing

De geschatte kosten voor aanleg, beheer en onderhoud zijn voor elk alternatief bepaald. De bouwkosten zijn berekend op basis van het handboek van het Thaise Bureau van de Begroting, gebaseerd op eenheidstarieven per april 2009 (n.b. het betreft ruwe ramingen). Naast de bouwkosten bevat deze raming ook de kosten die verband houden met beheer en onderhoud. Er is verondersteld dat deze kosten jaarlijks ongeveer 5% van de bouwkosten bedragen met het oog op de primaire doelstellingen van interventies tijdens hun levensduur die doorgaans gesteld wordt op 50 jaar. De te verwachten schade is ook bepaald op basis van deze levensduur. Bovendien moet worden opgemerkt dat in de geraamde aanlegkosten nog niet de kosten van grondverwerving zijn opgenomen, voor de terreinen waar de beoogde mitigerende maatregelen dienen te worden gerealiseerd. Daarom moeten de gepresenteerde kosten als indicatief worden beschouwd en is gedetailleerd onderzoek nodig om nauwkeuriger kostenramingen te verkrijgen, Deze zullen waarschijnlijk hoger uitvallen. Hierdoor is in de huidige studie de focus op grondgebruik, hydrologische en hydraulische aspecten geweest en heeft de kosten baten analyse een verkennend karakter gehad.

Optimaal niveau van mitigerende maatregelen tegen overstroming

De effecten van een overstroming kunnen worden beperkt, en daarmee kan het verlies aan mensenlevens en de materiële schade worden verminderd. Keuze van een bepaald alternatief ter vermindering van de kans op overstroming is afhankelijk van de hydrologische en hydraulische kenmerken van het riviersysteem. Maatregelen ter beperking van overstromingen kunnen echter niet vanuit een oogpunt worden beoordeeld. De technische prestaties van deze maatregelen, in termen van voorkoming van overstroming en de daaruit voortvloeiende schade, in de beschouwingen worden betrokken, want dit is belangrijk voor een algehele beoordeling van de aanvaardbaarheid van elk alternatief.

Het optimale niveau van beperking van overstromingen kan niet alle overstromingsrisico's elimineren. Normaal gesproken kan worden verwacht dat alleen de kosten ter beperking van overstromingen en schade door resterende overstromingen worden geminimaliseerd. Het betreft voor elke alternatieve oplossing voor het beperken van overstromingen het punt waar de som van de constructie, beheer en onderhoud kosten en schade worden geminimaliseerd.

Identificeren en beoordelen van de haalbaarheid van interventies ter beperking van overstromingen

Het stroomgebied van de Chi rivier is onderworpen aan overstromingen en verhoogde overstromingsrisico's voor mensen en bezittingen als gevolg van fysieke en operationele beperkingen van de bestaande hoogwaterbeheersing systemen, betrouwbaarheid van bestaande voorzieningen voor hoogwaterbeheersing die niet het vereiste niveau van bescherming bieden, de veranderingen in het grondgebruik in overstromingsgevoelige gebieden en de beperkte kennis van de overstromingsrisico's. Betrouwbare voorzieningen ter beperking van overstromingen zouden echter zorg moeten dragen voor een minimale verstoring en onderhevig moeten zijn aan acceptatie door de lokale bevolking. Efficiënt waterbeheer kan worden bereikt door ervoor te zorgen dat passende maatregelen ter beperking van overstromingen tijdig worden gerealiseerd met meer aandacht voor het type en de locatie van de maatregelen. Deze studie biedt een inventarisatie en beschrijving van de verschillende opties voor overstromingsbeheer in het stroomgebied van de Chi rivier. Alleen de meest voorkomende hedendaagse maatregelen voor een geïntegreerde aanpak van de hoogwaterbeheersing zullen worden besproken. Deze alternatieven zijn gekozen op basis van begrip van de fysieke situatie, analyse van het gedrag van overstromingen, identificatie van de behoeften en de evaluatie van eerdere studies. De alternatieve maatregelen moeten uiteindelijk voldoen aan de volgende criteria:

- haalbaarheid, met het oog op het in beschouwing nemen van verwachte overstromingsschade;
- technische effectiviteit, met het oog op de effectiviteit bij het verminderen van de omvang van overstromingen.

Voorafgaand aan het bepalen van de mogelijke maatregelen ter beperking van overstromingen is het hydraulische model toegepast om te bepalen welke gebieden gevoelig zouden kunnen zijn voor overstromingen met een kans van overschrijding van 1% per jaar, alsmede hoe de voorgestelde maatregelen zodanig kunnen worden toegepast dat dit niet zal leiden tot negatieve benedenstroomse effecten.

Om te komen tot de simulatie van een overstroming op basis van een scenario van een overstromingramp met een kans van overschrijding van 1% per jaar, zijn vier mogelijke alternatieven ter beperking van overstromingen geanalyseerd om de hydraulische effectiviteit te ramen bij het beperken van overstromingen op kritieke plaatsen in het stroomgebied van de Chi rivier. Mogelijke hydraulische effecten om overstromingen beperken zijn beoordeeld door het vergelijken van de diepte en mate van overstroming voor en na de maatregelen. Om de meest effectieve maatregelen te bepalen zijn indicatieve kosten-baten analyses uitgevoerd. Op basis hiervan is door het minimaliseren van de resulterende totale kosten het optimaal functioneren van de voorkeurs maatregel gevonden.

Rivier normalisatie

Overstromingen worden veroorzaakt door een afvoer die groter is dan de capaciteit van de bestaande rivier waardoor de aangrenzende landen overstromen. De afvoer capaciteit van de rivier is vaak ingeperkt ten gevolge van vernauwing door uitgebreide groei van vegetatie, ophoping van vuil, enz. Dit is de reden waarom rivier normalisatie een haalbare optie kan zijn en in sommige gevallen kan dit worden toegepast door de hydraulische ruwheid te verkleinen, door de dwars doorsnede te vergroten, of door de kans op verstoppingen en samenklontering van drijvend vuil te verminderen.

Er is een schatting gemaakt om de mogelijke effecten van rivier normalisatie op de afvoer functie te bepalen, die de top afvoer van de rivier zal verminderen. Daarom is rivier normalisatie vervolgens gemodelleerd door het veranderen van de hydraulische weerstand, dwz. de Manning ruwheid coëfficiënten van de rivierbedding, die waren bepaald op basis van het kalibratie proces, in het 1D/2D SOBEK hydraulische model.

Zeven scenario's voor rivier normalisatie werden gedefinieerd en geanalyseerd, die overeenkomen met het aantal secties over het gehele traject van de simulatie (te weten van waarnemingspunt (E5), Ban Non Puai tot aan de samenvloeiing van de Chi en de Mun rivier). Als de rivier normalisering van de Chi rivier nader zou moeten worden beschouwd dan zijn de volgende acties nodig:

- maak een voorlopige beoordeling van de kosten van normalisatie;
- kwantificeer de voordelen die kunnen worden verkregen door het normaliseren van de Chi rivier ter beperking van overstromingen tijdens een hoogwater met een kans van overschrijding van 1% per jaar.

Het 1D/2D SOBEK model van de Chi rivier heeft aangetoond dat het verwijderen van hydraulische obstakels, van vegetatie, en schoonmaken van de oevers in de benedenloop van de rivier, dat wil zeggen het normaliseren van de sectie van de Chi - Lam Pao samenvloeiing tot aan de Chi - Mun samenvloeiing, een significant effect zal hebben op de kenmerken van overstroming in de overstroombare gebieden. In het hydraulische model zijn de Manning n ruwheid coëfficiënten van de benedenloop vastgesteld op 0,032 (van de Chi - Lam Pao samenvloeiing tot aan de Chi - Nam Yang samenvloeiing) en 0,028 (van de Chi - Nam Yang samenvloeiing tot aan de Chi - Mun samenvloeiing). Ervan uitgaande dat de rivier normalisatie deze ruwheid coëfficiënten reduceert tot 0,025 dan zal het overstroomde gebied met een kans van overschrijding van 1% per jaar met ongeveer 16.000 ha afnemen. De kosten van schade in verband met wateroverlast (zowel fysische eigenschappen en landbouwgrond) zal naar verwachting over de hele stroomgebied afnemen van 86 miljoen US$ tot 64 miljoen US$. De resultaten tonen aan dat het voorkomen schade kan worden beschouwd als een voordeel van rivier normalisatie die de geraamde kosten van 21 miljoen US$ te boven gaat.

Het is noodzakelijk om zorgvuldig aandacht te schenken aan de keuze van de methode voor rivierverbetering, omdat deze een aantal belangrijke tekortkomingen heeft waarmee rekening moet worden gehouden. Rivierverbetering moet het bestaande morfologische evenwicht van het riviersysteem niet verstoren. Ter compensatie van de wijziging in de hydraulische variabele zullen, om een nieuw/stabiel evenwicht te bereiken, andere parameters zich aanpassen. Daarom zal rivierverbetering waarschijnlijk een grote invloed op een rivier hebben, omdat het bestaande fysieke evenwicht van de rivier en de mogelijk grote gevolgen daarvan voor het milieu (minder soorten, effect op de aanvulling van het grondwater, enz.) worden verstoord, tenzij de werkzaamheden zorgvuldig worden voorbereid, uitgevoerd en gecontroleerd. Ook moet regelmatig onderhoud worden uitgevoerd om een duurzame en bevredigende werking te garanderen.

Reservoir beheer

Reservoirs voor hoogwaterbeheersing kunnen in belangrijke mate bijdragen aan de beperking van overstromingen omdat ze afvoerpieken die stroomopwaarts van de overstroombare gebieden optreden op kunnen vangen en tijdelijk bergen, en later geleidelijk afvoeren. Grote stukken land die nodig zijn om een nieuw reservoir te vestigen zijn niet meer beschikbaar, met name omdat er grote bewoond gebieden zijn en goede landbouwgronden.

De bijdrage van de Nam Phong en Lam Pao rivieren, waar de Ubol Ratana en Lam Pao reservoirs al aanwezig zijn, aan beperking van overstromingen in het stroomafwaartse deel van het stroomgebied van de Chi rivier is bestudeerd. De gemodelleerde scenario's zijn onderzocht door het veranderen van het initiële reservoir volume en de dagelijkse afvoeren vanuit de bestaande reservoirs aan het begin van het regenseizoen, om de gevolgen van een overstroming met een kans van overschrijding van 1% per jaar te verminderen. De voorwaarden voor een optimaal ontwerp en beheer, te weten het percentage van het initiële reservoir volume en reservoir afvoer, zijn bepaald door het toepassen van de waterbalans vergelijking berekend op basis van instroom, uitstroom, verliezen, en berging in het reservoir. Bovendien zijn bij het bepalen van de reservoir afvoeren de benedenstroomse behoeften voor watervoorziening, landbouw en milieu, de eisen met betrekking tot overstromingen, overwegingen met betrekking tot opslag, en wettelijke vereisten voor de minimum afvoer ook in aanmerking genomen.

Simulatietechnieken zijn niet in staat om direct een optimale oplossing voor een alternatief reservoir beheer te genereren. Echter, door verschillende model simulaties met alternatieve beheer regels, kunnen vrijwel optimale operationele oplossingen worden geïdentificeerd. In dit onderzoek zijn vijf scenario's van reservoir beheer vergeleken om de meest efficiënte vorm van beheer voor vermindering van potentiële overstromingsschade te identificeren. De resultaten toonden aan dat het scenario waarbij tijdens het hoogwater seizoen, de afvoeren uit Ubol Ratana en Lam Pao reservoirs respectievelijk ongeveer 55% en 75% van hun oorspronkelijke dagelijkse afvoer zijn, terwijl de reservoir volumes aan het begin van de natte moesson ongeveer respectievelijk 50% en 30% van hun oorspronkelijke volume zijn de beste mogelijkheid biedt. Door de voorgestelde regels voor het reservoir beheer te volgen, is een reductie van ongeveer 14.100 ha in overstroomd gebied en zal de potentiële schade door overstromingen naar verwachting van 86 miljoen US$ tot 70 miljoen US$ dalen. Er moet worden opgemerkt dat voor het alternatieve reservoir beheer geen extra kosten voor aanleg, beheer en onderhoud nodig zijn, en dat alleen de regels voor het beheer van het reservoir moeten worden aangepast. Echter, voor het beheer van de reservoirs is een zeer zorgvuldige afweging in verdere studies nodig, met speciale aandacht voor de directe en indirecte schade, waaronder verlaging van levering van energie vanuit het reservoir, irrigatie, watervoorziening, toerisme, enz., die waarschijnlijk zullen leiden tot verhoging van de kosten voor dit alternatief. Voor beide reservoirs is echter een significante toename van de opslagcapaciteit tot het einde van het regenseizoen mogelijk om tijdens dreigende overstromingen water op te slaan. Bovendien moet men ermee rekening houden dat overschatting van het initiële reservoir volume en reservoir afvoer kan leiden tot substantiële verliezen. Daarnaast, kan uitsluitend reservoir beheer in de praktijk alleen doorgaans niet worden beschouwd als een betrouwbaar alternatief om enorme topafvoeren veilig te verwerken. Daarom zou dit alternatief moeten worden geïntegreerd met andere maatregelen ter beperking van overstromingen, bijvoorbeeld door middel van wetgeving en het reguleren van ontwikkelingen in het overstroombare gebied.

Groene rivier (nevengeul)

In het stroomgebied van de Chi rivier, overstromen de belangrijkste overstroombare gebieden regelmatig, wat wordt veroorzaakt door overtopping van laaggelegen punten op de oevers van de Chi rivier. Een mogelijke oplossing om de kans op overtopping te beperken, is om een reeks nevengeulen aan te leggen ten einde de afvoer binnen de capaciteit van de Chi-rivier te houden en daardoor de effecten van topafvoeren op het hoofdsysteem te beperken (N.B: het wordt 'groene rivier' genoemd, omdat de nevengeul in droge perioden kan zorgen voor een natuurlijke groene ruimte, zoals park of agrarisch gebied na de afvoer piek. In andere omstandigheden, kan het water in deze zone ook worden opgeslagen voor agrarische doeleinden).

Om de locaties voor een nevengeul te bepalen is hydraulische modellering uitgevoerd om het overstroomde gebied dat zou voortvloeien uit een overstroming die overeenkomt met een kans van overschrijding van 1% per jaar te bepalen. De gevonden nevengeul is gedefinieerd als de geul en die delen van het overstroombare gebied naast de geul, die redelijkerwijs nodig zouden zijn om te voorzien in een veilige passage van de piekafvoer. Voor de selectie van het meest veelbelovende scenario zijn zeven mogelijke scenario's geanalyseerd en geoptimaliseerd met het 1D/2D SOBEK model. Gebleken is dat het alternatief bestaande uit een systeem met twee nevengeulen over ongeveer 53 km op de rechteroever van de overstroombare gebieden van de Chi rivier het meest effectief lijkt te zijn. Het deel van de stroom dat groter is dan de capaciteit van de Chi rivier wordt omgeleid naar de nevengeulen. De eerste afleiding uit de Chi rivier komt in de buurt van Chi-Lam Pao samenvloeiing door middel van een 36 km lange nevengeul. Bovendien wordt bij topafvoeren water opnieuw afgeleid via een andere nevengeul, die over een lengte van 17 km zou moeten worden gelokaliseerd in het meest stroomafwaarts gelegen einde van de Chi rivier. Het voorontwerp van de nevengeul is gemaakt op basis van een trapeziumvormige geul met een bodem breedte van 25 m, taluds 1:1 en een diepte van 3 m. De controle zou moeten plaatsvinden in de vorm van stuwen of schuiven bij de ingang van de nevengeul. Met deze nevengeulen, kan het overstroomde gebied met een kans van overschrijding van 1% per jaar met 8000 ha worden verminderd. Bovendien zouden de mogelijke voordelen van maatregel, die worden gedefinieerd als de afname van de schade kosten, ongeveer 10,8 miljoen US$ bedragen. Aangezien de kosten voor aanleg, beheer en onderhoud van dit alternatief worden geschat op ongeveer 9,3 miljoen US$, kan deze maatregel nog steeds als gunstig worden beschouwd, indien geen rekening wordt gehouden met de kosten voor grondverwerving voor de nevengeulen. Echter, het grondgebruik binnen de grenzen van de nevengeulen is aan beperkingen onderhevig om de vrije afvoer van hoogwater of de veiligheid niet in gevaar te brengen. In het algemeen worden activiteiten zoals landbouw, begrazing, en wetland toegestaan, mits een goed waarschuwingssysteem voor overstromingen op zijn plaats is, omdat ze snel en gemakkelijk verwijderd kunnen worden of dat zij weinig beperking voor de afvoer opleveren.

Retentiebekkens

In antwoord op de onvermijdelijke overstromingen, is het realiseren van opslag buiten het rivierbed een mogelijke maatregel om door vermindering van overstromingen door overtopping de daarmee samenhangende schade te beperken. Met het opslag bekken kan worden bewerkstelligd dat het opslagbekken onder gecontroleerde omstandigheden wordt gevuld bij waterstanden en afvoeren boven een bepaalde grens. Een inlaatwerk, dat op basis van vooraf gedefinieerde voorwaarden wordt bediend, zal worden ingezet om een deel van het hoge water te in het opslagbekken te bergen om de diepte en mate van de topafvoer verder stroomafwaarts te verminderen. Het water in het opslagbekken wordt afgelaten wanneer de Chi rivier de stroom kan verwerken zonder dat het water hoger komt dan het niveau van de oevers. Omdat het gewenste effect van een opslagbekken stroomafwaarts van het stroomgebied van de Chi rivier optreedt, heeft een zo vroeg mogelijke locatie, waarbij toch al grote hoeveelheden water kunnen worden opgeslagen, de voorkeur.

In deze studie zijn in de voorbereidende fase vier mogelijke locaties gelegen langs de Chi rivier in aanmerking genomen. Vervolgens is de meest veelbelovende locatie geselecteerd voor een meer gedetailleerde analyse. Uit de resultaten bleek dat met een opslagbekken in het bovenstroomse deel van het stroomgebied van de Chi rivier stroomafwaarts een aanzienlijk dempend effect kan worden verkregen. Om deze reden is gekozen voor de aanleg van 5.500 ha stroomopwaarts gelegen waterberging capaciteit op de linkeroever van de overstroombare gebieden van de Chi rivier, gelegen tussen de waarnemingspunten Ban Khai en Ban Kaeng Ko, die doorgaans worden beschouwd als herhaaldelijk overstroomde gebieden. Om de vereiste opslag capaciteit te realiseren, zou voor het opslagbekken, de bouw van een dijk nodig zijn met een maximale hoogte van 9 m, trapeziumvormige nevengeulen (bijvoorbeeld 25 m breed op de bodem, taluds 1:1 en 3 m diep), en inlaatwerken om het mogelijk te maken bepaalde delen van het overstroombare gebied tot een specifieke diepte op bepaalde tijdstippen te inunderen. Door de het leiden van de ontwerp afvoer door het potentiële retentiebekken, werd een vermindering van ongeveer 5.000 ha in het overstroomde gebied gerealiseerd. Uit een kosten-baten analyse bleek dat voor een investering van ongeveer 17,0 miljoen US$, de geselecteerde toepassing van een opslagbekken door het verminderen van overstromingsschade een voordeel van 17,3 miljoen US$ zou opleveren.

Selectie van alternatieve maatregelen

Om te komen tot een keuze van de set mitigerende maatregelen tegen overstromingen in deze studie zou moeten worden beschouwd, zijn de alternatieve maatregelen door middel van een selectie proces op basis van een beslissing matrix vergeleken. De analyse van de bovengenoemde resultaten blijkt dat de alternatieven voldoen aan de selectiecriteria, en inderdaad veelbelovend genoeg zijn om verder te worden overwogen. Omdat echter enkele alternatieven soms leiden tot negatieve effecten op andere alternatieven, zou een geschikte combinatie wenselijk zijn om het totale risico te beperken.

Optimale combinatie van maatregelen voor het mitigeren van overstromingen

Het is niet mogelijk om overstromingen te voorkomen of zelfs overal te verminderen. Daarom is het noodzakelijk om het meest effectieve gebruik te maken van middelen en kosten. In het kader van de groeiende interesse in het vinden van passende oplossingen voor het beheer van de overstromingsrisico's in het stroomgebied van de Chi rivier, kunnen doeltreffende antwoorden leiden tot een set of oordeelkundige combinatie van maatregelen ter beperking van overstromingen in plaats van afhankelijkheid van een enkele maatregel. In deze studie, van de beste case scenario voor elk alternatief, zijn een aantal scenario's geformuleerd, te weten een totaal van 11 mogelijke combinaties / scenario's zijn als volgt geanalyseerd (N.B: de totale kosten voor aanleg, beheer en onderhoud variëren in elk scenario):

- Scenario I-VI: combinatie van twee alternatieven;
- Scenario VII-X: combinatie van drie alternatieven;
- Scenario XI: combinatie van alle vier de alternatieven.

Door middel van gedetailleerde hydraulische modellering, is het meest effectieve scenario bepaald door middel van een reeks prioritaire punten op grond van technische en efficiëntie aspecten. Uit de onderzoeken bleek dat Scenario XI als beste naar voren kwam, dat de volgende belangrijkste maatregelen omvat:

- over ongeveer 331 km normalisatie van de rivier;
- veranderingen in de beheerregels voor de Ubol Ratana en Lam Pao reservoirs;
- twee nevengeulen, in totaal 53 km lang;
- 5.500 ha opslagbekkens.

Uit zowel technisch als financieel oogpunt, zou de optimale combinatie leiden tot een grotere afname van de schade dan een van de individuele maatregelen, de afname van de overstromingsschade kan 48 miljoen US$ bedragen. Bij de huidige benadering is de potentiële schade, die kan worden ondervangen door toepassing van de optimale combinatie van maatregelen ter beperking van overstromingen groter is dan de kosten voor aanleg, beheer en onderhoud, dat wil zeggen 47 miljoen US$. In dit verband is het opmerkelijk dat een optimale oplossing ertoe zou leiden dat het overstroomde gebied bij voor een overstroming met een kans van overschrijding van 1% per jaar wordt teruggebracht van circa 143.000 ha tot 101.000 ha.

Interacties tussen ruimtelijke ordening en overstromingen

Een substantieel deel van het stroomgebied van de Chi rivier is voornamelijk veranderd als gevolg van verhoogde druk door menselijke ingrepen voor bewoning, uitbreiding van de landbouw, ontwikkeling van infrastructuur, ontbossing, enz. De veranderingen in grondgebruik beïnvloeden het hydrologische gedrag van een stroomgebied door de invloed van grondgebruik op de generatie van afvoer processen. Deze veranderingen kunnen leiden tot wijzigingen van de hoeveelheid oppervlak/generatie van grondwater stroming, overstroming regimes en omvang. Om de gevoeligheid voor overstromingen in het stroomgebied van de Chi rivier te beperken, maakt de Koninklijke Thaise Regering het mogelijk om belangrijke veranderingen aan te brengen teneinde te komen tot een beter beheer in de landbouw, bosbouw, ruimtelijke ordening, waterbeheer, en verstedelijking. Om het effect voor het hele stroomgebied te kwantificeren, is allereerst een evaluatie van de veranderingen in grondgebruik geanalyseerd.

De ontwikkeling van potentiële scenario's voor veranderingen in het stroomgebied weerspiegelen mogelijke toekomst scenario's over 50 jaar. Hoewel het niet mogelijk is om te onderkennen wat er precies zal gebeuren in de komende 50 jaar, kunnen toekomstige trends worden geprojecteerd op de schaal van de verandering die in overstromingsrisico's zou kunnen optreden. Volgens het door de regering opgestelde plan voor de ruimtelijke ordening in 2057, zullen veranderingen in het grondgebruik naar verwachting worden gedomineerd door bebossing. Het beboste deel van het totale stroomgebied zal naar verwachting toenemen van 17,9% in 2000-2002 tot 37,2% in 2057. De toename in beboste gebieden zal resulteren in afnamen in alle niet bebost grondgebruik. Agrarische gebieden (75,5% van het totale stroomgebied) zullen met de grootste oppervlakte afnemen (tot 18,9% in 2057). De stedelijke gebieden zullen met 0,9% afnemen en deze afname zal worden vervangen door de industrie voor ongeveer 0,5%.

Individuele model simulaties hebben de mogelijkheden van het hydrologische model aangetoond om de gevolgen van veranderingen in grondgebruik op het hydrologische regime van het stroomgebied van de Chi rivier te bepalen. De simulatieresultaten geven aan dat de toekomstige veranderingen in grondgebruik zullen leiden tot een significante afname van piekafvoeren. Deze kunnen aanzienlijk afnemen waardoor de potentiële bedreigingen van overstromingen in het bovenstroomse deel van het stroomgebied van de Chi rivier in vergelijking met het huidige grond gebruik als een systematische en realistische benadering van bebossing in de ruimtelijke ordening in het stroomgebied van Chi rivier wordt toegepast. Zo kan de afname van de topafvoer (met een kans van overschrijding van 1% per jaar) bij het afvoerpunt van een van de bovenstroomse deelstroomgebieden leiden tot een afvoer met een kans van overschrijding van ongeveer 10% per jaar. Echter, de verschillen in topafvoeren en afvoer volume variëren sterk van het ene deelstroomgebied naar het andere.

Bij de analyse van het toekomstige grondgebruik in het stroomgebied van de Chi rivier is geconstateerd dat bij veel van de locaties die momenteel worden getroffen door overstromingen, het niet waarschijnlijk lijkt dat zulke overstromingen in de toekomst in betekenende mate zullen afnemen. Met andere woorden, het aanpassen van de ruimtelijke ordening zal geen belangrijk reducerend effect hebben op het terugdringen van overstromingen in de benedenstroomse gebieden, omdat de reducties in the afvoer vanuit de deelstroomgebieden klein zijn. Een onderliggende veronderstelling is dat het klimaat in deze regio stabiel blijft en dat alle andere mogelijke veranderingen in het milieu worden beschouwd in de model benaderingen.

Om beter te begrijpen hoe veranderingen in het grondgebruik toekomstige overstromingsrisico's kunnen beïnvloeden, is het 1D/2D SOBEK model gebruikt om toekomstige veranderingen tegen de huidige resultaten te toetsen. De resultaten toonden aan dat tijdens een overstroming met een kans van overschrijding van 1% per jaar de totale oppervlakte aan overstromingen over het stroomgebied als geheel slechts in geringe mate zal afnemen, dus van 143.000 ha naar 142.000 ha. De afname van de omvang van de overstroming is daarom marginaal in het kader van het toekomstige grondgebruik scenario. Hoewel de omvang van de overstromingen in de toekomst waarschijnlijk licht zal afnemen zullen de kosten van schade door overstromingen aanzienlijk blijven stijgen als gevolg van de meest opmerkelijke verandering in het aantal commerciële eigendommen in de overstroombare gebieden. In het kader van het toekomstige grondgebruik scenario, zullen de kosten van de schade verbonden aan een overstroming over het hele stroomgebied naar verwachting stijgen van 86 miljoen US$ tot 140 miljoen US$. Ten einde de stijging van de kosten van overstromingsschade te minimaliseren is een scenario gebruikt om de effectiviteit van de optimale combinatie van overstroming beperkende maatregelen tegen een reeks van waarschijnlijke toekomstige situaties te testen. De resultaten suggereren dat het gebruik van de gekozen optimale oplossing kan helpen om door toekomstige overstromingen veroorzaakte schade te verminderen van 140 miljoen US$ tot 85 miljoen US$, en het overstroomde gebied temet nog 39.000 ha te verkleinen.

In verband met de betrouwbaarheid van de met het model gesimuleerde gevolgen van veranderingen in grondgebruik, is het essentieel om onzekerheid in de ruimtelijke ordening gegevens zoveel mogelijk uit te filteren. In praktijk toepassingen, zal altijd onzekerheid bestaan in model resultaten als gevolg van onzekerheid over de ruimtelijke verdeling van de voorspelde veranderingen in grondgebruik, in aanvulling op algemene model onzekerheden zoals die van de andere invoer gegevens, de structuur van het model en de parameters. Echter, vanwege het feit dat de onzekerheid analyse van grondgebruik moeilijk te beoordelen is en kan worden toegeschreven aan een aantal oorzaken, dient het verkleinen van deze onzekerheden te worden behandeld in nader onderzoek.

Slot opmerkingen

Het stroomgebied van de Chi rivier vormt ongeveer een derde van de landmassa en de bevolking van de noordoostelijke regio van Thailand. In relatie tot de geografie kan worden gezegd dat het landschap in feite een glooiend hoogland is met ongunstige agrarische omstandigheden gekenmerkt door onvruchtbare zandgronden met uitzondering van de overstroombare gebieden. Het neerslag regime is heel onregelmatig en onbetrouwbaar, waardoor het gebied kwetsbaar is voor droogte en overstromingen. De ontwikkelingen tot nu toe zijn voornamelijk bepaald door de uitbreiding van de directe exploitatie van natuurlijke rijkdommen. Landbouwsystemen, die volledige benutting van land en water met zich meebrachten, leidden onvermijdelijk leiden tot achteruitgang in natuurlijke hulpbronnen en verspreidde zich steeds meer.

Door de bevolkingsgroei in het stroomgebied van de Chi rivier is sprake van een enorme mate van ontbossing. Tussen 1952 en 2003 is het bos areaal met 20% afgenomen, terwijl de landbouw- en stedelijke gebieden snel in omvang zijn toegenomen. Het gebied wordt met name gekenmerkt door agrarisch grondgebruik, ongeveer 78% van de huishoudens zijn betrokken bij de landbouw, en zijn in sterke mate afhankelijk van een aantal gewassen, te weten rijst, samen met cassave, suikerriet, maïs, soja, pindas en kenaf. De huidige praktijk leidt echter tot erosie, verzilting en verlies van vruchtbaarheid, als gevolg van uitbreiding van het telen van cassave, suikerriet en maïs, die allemaal locatie afhankelijk lijken te zijn. Uiteraard kan men zien dat gewassen tegenwoordig meer divers zijn dan vroeger het geval was, en ze worden steeds vaker geteeld op geschikte gronden in het gebied, mede door resultaten in de ontwikkeling van de landbouw. Momenteel is er ongeveer 0,54 miljoen ha of ongeveer 11% van het totale areaal geïrrigeerd in het kader van 1.836 grote, middelgrote en kleine programma's op het gebied van waterbeheer, met inbegrip van het leveren van elektrische pompen. Bovendien is de industrie ook in toenemende mate van belang, aangezien het vooral ondersteunende diensten levert die een essentieel onderdeel vormen van de agrarische productie, die rijst, tapioca en suiker molens omvat. Als gevolg van de ontwikkeling van activiteiten in het stroomgebied van de Chi rivier, wordt onderkend dat de economie in de afgelopen decennia sterk is gegroeid.

In zekere mate, komen ontwikkelingen in het verleden niet noodzakelijkerwijs overeen met die in de toekomst. Daarom zou rekening moeten worden gehouden met de verwachte toekomstige effecten van de activiteiten, omdat ze waarschijnlijk zullen leiden tot een significante negatieve invloed op de functies in het stroomgebied en hun gedrag. In de toekomst zal de voortdurende groei en ontwikkeling in het stroomgebied van de Chi rivier nog steeds gebaseerd zijn op de exploitatie van natuurlijke hulpbronnen. Er moet echter met zorg worden omgegaan om de harmonie te bewaren tussen duurzame bronnen en de sociaal-economische eisen, maar ook om ervoor te zorgen dat de productiviteit van de primaire natuurlijke hulpbronnen niet zal verslechteren als gevolg van de ontwikkeling van activiteiten in het stroomgebied. In de omgang met de met grondgebruik samenhangende hulpbronnen, is optimalisatie van het grondgebruik essentieel gevonden voor het bereiken van economische en sociale voordelen. In het bijzonder in de agrarische sector, zijn paradigma verschuivingen nodig van extensieve naar intensieve landbouwsystemen, evenals van de afhankelijkheid van een paar marktgewassen tot meer diverse gewassen in combinatie met efficiënter gebruik van beperkte middelen. In lijn met de landbouw strategie, is de aanleg en de verbetering van irrigatie faciliteiten om goed te functioneren en goed te zijn aangepast aan de plaatselijke omstandigheden ook van belang met het oog op de verbetering de productiviteit van het watergebruik wat bijdraagt aan duurzaamheid op lange termijn van de landbouw. Kortom, de essentie van integrale planning is dat hierbij zoveel mogelijk rekening moet worden gehouden met de te verwachten ontwikkelingen, in samenhang met zowel schadelijke neveneffecten en nog onbekende voordelen van ontwikkelingsmaatregelen en manieren moeten worden gevonden om de voordelen te maximaliseren en de negatieve effecten zoveel mogelijk te beperken door efficiënt beheer.

Wanneer ontwikkeling van een stroomgebied plaatsvindt is het van essentieel belang om bewust te zijn van de wijze waarop de ontwikkeling kan worden geassocieerd met de mogelijkheid van het genereren van overstromingen, in de zin dat de ontwikkeling het overstromingsproces en vice-versa zou kunnen beïnvloeden. De veranderingen in het grondgebruik in verband met de ontwikkeling van het stroomgebied zouden moeten worden beschouwd, omdat er sprake kan zijn van het concentreren en versnellen van de afvoer, wat kan leiden tot meer overmatige schade aan de bestaande infrastructuur. De specifieke informatie die in dit verband relevant is betreft het overstroomde gebied en de kosten van overstromingsschade door een overstroming met een kans van overschrijding van 1% per jaar. Op basis van het huidige grondgebruik toont het afgeleide resultaat aan dat een oppervlakte van 143.000 ha zal worden overstroomd met een geschatte schade van 86 miljoen US$. De potentiële schade zou toe kunnen nemen tot 140 miljoen US$, ofwel 1,6-voudige als gevolg van veranderingen in de patronen van grondgebruik in de periode tot 2057.

In een poging om het functioneren van het stroomgebied te verbeteren, is het dus noodzakelijk om bijzondere aandacht te besteden aan het in balans brengen van de ontwikkeling voorwaarden voor ontwikkeling en verliezen in verband met overstromingen. Daarom is een breed scala van mogelijke maatregelen ter beperking van overstromingen onderzocht. De resultaten van de selectie gaven aan dat het meest rendabele plan waarschijnlijk zal bestaan uit de combinatie van normalisatie van de rivier, reservoirbeheer, nevengeulen, en opslagbekkens. Zodra ze in samenhang worden toegepast neemt, gebaseerd op de verwachte omvang van de topafvoer met een kans van overschrijding van 1% per jaar, de overstromingsschade (bij de huidige patronen van grondgebruik) af in de orde van 48 miljoen US$, wat overeenkomt met een reductie van 42.000 ha van het overstroomde gebied. De afname in de toekomstige overstromingsschade zal naar verwachting 54 miljoen US$ zijn (afname van 39.000 ha van de overstroomde gebieden), als de verandering in grondgebruik optreedt zoals deze momenteel wordt voorspeld (tot 2057).

Naast een overstroming met een kans van overschrijding van 1% per jaar, is het vermeldenswaard dat de omvang van deze overstroming kan worden overschreden. Dit zou betekenen dat voor een verdere beoordeling van de toekomstige overstromingsrisico's anticipatie van cruciaal belang wordt. Voor wat dit aspect betreft, is een inzicht in een uitzonderlijk grote overstroming (bijvoorbeeld een overstroming met een kans van overschrijding van 0,1% per jaar) nodig, om het hogere niveau van veiligheid te positioneren. Indien een dergelijke overstroming, onder de huidige ruimtelijke omstandigheden, zich zou voordoen zou 165.000 ha worden overstroomd en dit kan leiden tot 10 miljoen US$ aan directe schade. Als de voorspelde veranderingen in grondgebruik plaatsvinden over een langere periode tot 2057, is de omvang van de overstroomde gebieden bijna gelijk aan die onder de bestaande omstandigheden van grondgebruik, maar de schade kan toenemen met zoveel als 18 miljoen US$. In reactie op de dreiging van deze intensivering van overstromingen naast de geïntegreerde oplossingen voor de mitigatie ervan, als de huidige tendensen in grondgebruik blijven, kan de reductie van het overstroomde gebied worden geraamd op 34.000 ha en kan de schade met 4 miljoen US$ verminderen. Inmiddels wordt bij de grondgebruik omstandigheden voor 2057 een reductie van de overstroming met 32.000 ha verwacht en een afname van de mogelijke schade van 6 miljoen US$. In dit verband zal het duidelijk zijn dat de optimale combinatie van maatregelen overstromingsschade en overstromingsrisico's in het stroomgebied van de Chi rivier aanzienlijk zou verminderen, maar dat de gevolgen van overstromingen nog steeds niet volledig worden geëlimineerd.

Kortom, het geïntegreerde model waarbij de hydrologische (SWAT) en hydraulische (1D/2D SOBEK) modellen zijn gekoppeld heeft aangetoond dat het gebruikt kan worden voor veel meer dan het opsporen, vaststellen, en het schatten van de omvang van overstromingen, schade en effecten. In het bijzonder kan het worden gebruikt om de gevolgen van veranderingen in grondgebruik in het hele stroomgebied te onderzoeken. Op basis van de bovenstaande uitspraken kan een goed begrip van de relatieve omvang en de richting van toekomstige veranderingen worden verkregen, die kunnen helpen om toekomstige overstromingsrisico's en schade te minimaliseren Dit kan gebruikt worden als de best beschikbare informatie om de hoogwaterbeheersing te verbeteren.

About the author

Kittiwet Kuntiyawichai was born on 10th July 1977 in Buriram, Thailand. He followed his primary and secondary school in Buriram. He received a BEng degree in Transportation Engineering from Suranaree University of Technology in Thailand in April 1999. From 1999 till 2001, he worked as lecturer in the Department of Civil Engineering, Buriram Technical College. During that first year, he continued his study in Graduate Diploma Programme in Teaching Profession (Teacher Professional Certificate, 1 year course) at Buriram Rajabhat University.

In May 2001, he continued his study in the Master's programme in Water Resources Engineering, Department of Civil Engineering, Faculty of Engineering, Khon Kaen University, Thailand, and in April 2004 he received his MEng degree with the thesis entitled 'An application of INFOWORKS RS model for flood routing in Chi River Basin'.

From July 2004 till November 2005, he worked as a lecturer in the Department of Civil Engineering, Faculty of Engineering, Khon Kaen University, Khon Kaen, Thailand, and he still holds this position until now. At the university, his main tasks include teaching, research and consultancy activities.

In November 2005, he joined the Hydraulic Engineering - Land and Water Development (HELWD) Core, Department of Water Engineering, UNESCO-IHE Institute for Water Education, Delft, the Netherlands, for his PhD study. During this time period, Mr. Kittiwet Kuntiyawichai attended the 5th Annual Mekong Flood Forum, May 2007, Ho Chi Minh City, Vietnam. In the same year, he gave a refereed poster presentation at the 4th International SWAT Conference, Delft, the Netherlands. Moreover, he presented papers in five international conferences regarding flood-related issues that were held in 2008, 2009 and 2010 in Canada, Finland/Estonia, Thailand, Lao PDR and Japan. In particular, he has been awarded the 'Best Presentation and Best Paper Award' for his talk and paper entitled 'Comprehensive flood mitigation and management in the Chi River Basin, Thailand' at the 7th International Symposium on Lowland Technology, which was held from 16 - 18 September 2010, Saga, Japan. In addition, he published two papers in international peer-reviewed journals and five symposium papers. Most of them were related to his PhD topic. Two more papers are currently under preparation.

For Product Safety Concerns and Information please contact our EU representative GPSR@taylorandfrancis.com Taylor & Francis Verlag GmbH, Kaufingerstraße 24, 80331 München, Germany

T - #0140 - 160425 - C76 - 244/170/15 - PB - 9780415631242 - Gloss Lamination